AF777013

Luminescence - OLED Technology and Applications

Edited by Sergei Pyshkin

Published in London, United Kingdom

IntechOpen

Supporting open minds since 2005

Luminescence - OLED Technology and Applications
http://dx.doi.org/10.5772/intechopen.78903
Edited by Sergei Pyshkin

Contributors
Abhishek Kumar Soni, Bheeshma Pratap Singh, Zhaofeng Wang, Fu Wang, Volodymyr Halyan, Inna Ivashchenko, Christopher Campbell, Joel Abrahamson, Hollis Beagi, Fay Salmon, Jang Jin Yoo, Dong Woo Kang, Jang Un Kwon, Sunho Park, Jeong Sub Hwang, DonGyou Lee, Bu Yeol Lee, In Byeong Kang, Alhama Arjona-Esteban, Barbara Szafranowska, Julian Ochsmann

First published in London, United Kingdom, 2020 by IntechOpen
IntechOpen is the global imprint of INTECHOPEN LIMITED, registered in England and Wales, registration number: 11086078, 7th floor, 10 Lower Thames Street, London, EC3R 6AF, United Kingdom
Printed in Croatia

British Library Cataloguing-in-Publication Data
A catalogue record for this book is available from the British Library

Additional hard and PDF copies can be obtained from orders@intechopen.com

Luminescence - OLED Technology and Applications
Edited by Sergei Pyshkin
p. cm.
Print ISBN 978-1-78984-131-2
Online ISBN 978-1-78984-132-9
eBook (PDF) ISBN 978-1-83881-014-6

Meet the editor

Professor Sergei L. Pyshkin, DSc, is Scientific Advisor of the Institute of Applied Physics, Academy of Sciences of Moldova, and Adjunct Professor and Senior Fellow of Clemson University, SC, USA. He is also a member of the US Minerals, Metals, & Materials Society (TMS) and was awarded the state prize of the Republic of Moldova for investigations in solid-state physics and microelectronics. His works deal with non-linear optics (multi-quantum absorption), electron and phonon transport phenomena, photoconductivity and light scattering, luminescence, crystal and thin film growth, molecular beam- and laser-assisted epitaxies, nanotechnology, lasers for medicine, and scientific instrument making (boxcar integrators and solid-state IR matrix photoreceivers). Dr. Pyshkin is included in Marquis' "Who's Who in America" (2008–2013) and "Who's Who in the World" (2009–present).

Contents

Preface

This edited volume is a collection of reviewed and relevant research chapters concerning developments within luminescence technology and applications. It includes contributions by various authors and is edited by a group of experts in organic light-emitting diode (OLED) technology.

The book includes the following chapters:

Chapter 1: Luminescent Materials in Lighting, Display, Solar Cell, Sensing, and Biomedical Applications

Chapter 2: Triboluminescence: Materials, Properties, and Applications

Chapter 3: Mechanism of Photoluminescence in Erbium-Doped Chalcogenide

Chapter 4: Optically Clear Adhesives for OLED

Chapter 5: Picture Quality and Sound Quality of OLED TVs

Chapter 6: TADF Technology for Efficient Blue OLEDs: Status and Challenges from an Industrial Point of View

The target audience comprises scholars and specialists in the field.

IntechOpen

Chapter 1

Luminescent Materials in Lighting, Display, Solar Cell, Sensing, and Biomedical Applications

Abhishek Kumar Soni and Bheeshma Pratap Singh

Abstract

This chapter comprises a broader extent of the luminescence phenomenon with the mechanism involved therein as well as applications. Typically, the up and down conversion and downshifting behavior of the optical materials have been elucidated in brief. The fundamental understanding of these optical materials has been described by using schematic representations. It is well documented that the rare earth-based optical materials are known for their luminescent enrichment due to availability of the ladder-like energy levels. These energy levels can be utilized for the excitation of the luminescent materials by using a suitable excitation source. In the process of development of luminescent materials, choice of host matrices and dopant ions is very crucial. Strong correlation of these optical materials has been shown with the current scenario of our society and daily life. In view of the ongoing research, nanophosphor, glasses, and quantum dots with size- and shape-dependent optical behavior have been given in detail. The involved mechanism and the energy transfer phenomenon have been well elucidated by schematic and figures for the evident explanation to the readers. Our emphasis is to elucidate these optical materials in the development of innovative multifunctional applications such as lighting, display, sensing, LEDs, solar cell, and biological applications.

Keywords: light conversion, phosphor, quantum dots, optical temperature sensing, LEDs, solar cell, biological and clinical

1. Introduction

1.1 Luminescence in rare earth ions

In 1888, German physicist Wiedemann for the very first time employed the term "luminescenz." Luminescence promotes emission of light from a material at a certain excitation wavelength. It can be categorized into diverse kinds depending upon the various involved processes of excitation. It is classified as photoluminescence, thermoluminescence, electroluminescence, and chemiluminescence. Photoluminescence is a process in which the emission of light (photon) can be obtained from the object upon excitation. It is subgrouped into fluorescence and phosphorescence. In the thermoluminescence phenomenon, emission of light from a solid material occurs upon heating. Electroluminescence arises from gases under applied electrical potentials, while chemiluminescence may prevail during the chemical reaction process [1–3]. The fluorescence phenomenon exhibits a short time duration of decay levels, while phosphorescence exhibits prolonged decay time after the seize of the excitation source. Inorganic materials accommodated with impurity foreign ions most commonly lanthanide ions exhibiting luminescence upon excitation are known as phosphors.

In 1994, Bhargava et al. [4] reported for the very first time on Mn^{2+}-doped ZnS DCNCs that exhibit characteristic features such as change in band gap and shifting of emission and excitation spectra of prepared nanomaterials. An external quantum efficiency of up to 18% was observed. This report opened the new pathways in the field of nanophosphor research area.

Optical materials doped with lanthanide activators are very promising candidates for photoluminescence studies. The luminescent material exhibiting luminescence properties can be utilized for solid state lighting (SSL), light emitting diodes (LEDs), display devices, solar cell, sensing, and biomedical applications. The advantage of these luminescent materials in diverse application fetches among the researchers to search the novel improved materials with magnified luminescence properties. RE ions exhibit ladder-type energy levels and sharp emission lines. Upon doping into a low phonon frequency host lattice, these ions can be easily excited by employing different excitation sources.

Up-conversion is a non-linear anti-stokes optical process in which a low energy input photon is converted into a high energy output photon by the process of sequential multiphoton absorption. Surprisingly, due to different involved energy transfers in different RE ions, the emission properties can be manipulated by suitable doping and/or co-doping.

This chapter has been considered to account for basic understanding involved in the luminescent materials, viz. up- and down-conversion and downshifting behavior of the optical materials. These nanomaterials could be a reasonable alternative for novel multi-functional optical materials utilized in lighting, display, sensing, LED, solar cell, biological, and clinical applications. Quantum dots (QDs) and their uses in the biological and clinical application have been justified in view of current uses. **Figure 1** shows the various applications of luminescent materials.

1.2 Rare earth elements

Commonly known as 4f- block elements having an atomic number ranging from 57 (La) to 71 (Lu), these elements adjusted themselves at the bottom of the Mendeleev periodic table.

1.3 Silent features of the rare earth elements

The f-block elements reveal various well-defined features that distinguish them from the d-block metals. Some of the peculiar properties of f-block elements are as follows:

i. The f-block elements manifest a range of coordination numbers (6–12).

ii. RE ions form ionic complexes that undergo facile exchange of ligands.

iii. RE ions reveal small and very sharp electronic spectra and less crystal-field splitting.

iv. Since 4f orbitals in the RE ions are well shielded by $5s^2$ and $5p^6$ orbitals, therefore their optical properties are mostly not changed by the host.

v. RE ions most commonly exist in its 3+ stable oxidation state.

vi. Some RE elements also reveal other valence states such as 2+ and 4+ under specific conditions.

1.4 Spectroscopy of rare earth elements

A partially filled 4f shell is well shielded from completely filled outer $5s^2$ and $5p^6$ orbitals which gives it sharp electronic spectra. The energy level spectra of REs do not alter to the outer environment in which they are indulged.

In general, lanthanide ions exist in 3+ state and also exhibit 2+ oxidation state in some case. RE ions in triplet states demonstrate intra-4f-4f transitions which thus result in an intense narrow emission band in a vast variety of host matrices. The shielding of 4f orbitals furnished by the $5s^2$ and $5p^6$ electrons promotes to radiative transitions which match well with the free ions. Some other RE ions such as Sm^{2+} and Eu^{2+} show the divalent character. These ions also reveal the visible luminescence under suitable excitation wavelength. Moreover, this book chapter is predominantly based on the luminescence of the trivalent RE ions.

Neutral RE elements reveal the electronic configuration [Xe] $4f^N 6s^2$ or [Xe] $4f^{N-1}$ $5d\ 6s^2$, where [Xe] constitutes a xenon core whose electronic structure is represented as [Xe] = $[1s^2 2s^2 2p^6 3s^2 3p^6 3d^{10} 4s^2 4p^6 4d^{10} 5s^2 5p^6]$. The electrostatic shielding of 4f orbitals by the outermost $5s^2 5p^6$ orbitals leads an atom-like behavior in RE ions under a solid host matrix such as a crystal or glass.

Term and symbol is represented as $^{(2S+1)}L_J$, where S, L, and J are total multiplicity, orbital angular momentum and angular momentum, respectively. Involved selection rules for intra f-f transitions in lanthanide ions are given in **Table 1**.

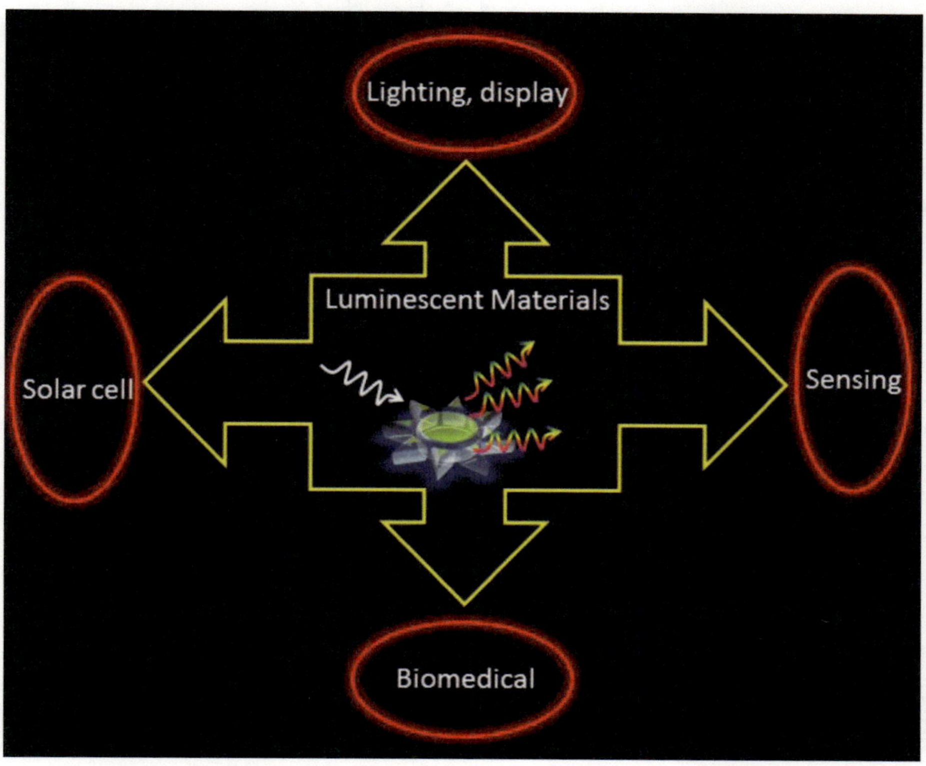

Figure 1.
Various applications of luminescent materials.

Operator	Parity	ΔS	ΔL	ΔJ*
Electric dipole	Opposite	0	≤6	≤6 (2, 4, 6, if J or J′ = 0)
Magnetic dipole	Same	0	0	0, ±1
Electric quadrupole	Same	0	0, ±1, ±2	0, ±1, ±2

**J = 0 to J′ = 0 transitions are forbidden always; J = 0 → J′ = 1, 3, 5 exhibit weak transition intensity; and J = 0 → J′ = 2, 4, 6 convey strong transition intensity.*

Table 1.
Selection rules for intra f-f transitions in lanthanide ions.

1.5 Rare earth ion levels in solids

According to Laporte rule, 4f-4f transitions are forbidden and show a very weak oscillator strength of about ~10^{-6}.

The Hamiltonian of a particular RE ion can be represented as follows [5]:

$$H = H_{free\ ion} + V_{Host} + V_{EM} + V_{Vib.} + V_{ion-ion} \quad (1)$$

where $H_{free\ ion}$ represents the Hamiltonian of an isolated ion, V_{Host} describes the environment effect on the RE ion, V_{EM} describes RE ion with an electromagnetic field interaction, V_{Vib} describes the interaction of RE ion with host lattice vibrations, and $V_{ion\text{-}ion}$ describes the interaction between the RE ions, respectively. The Hamiltonian of free ion ($H_{free\ ion}$) can be splitted into several components, i.e.,

$$H_{free\ ion} = H_{CF} + H_C + H_{SO} \quad (2)$$

where H_{CF} belongs to central-field Hamiltonian which reveals the interaction of an orbiting electron with the nucleus and effective field from all other electrons, while H_c term represents the residual Columbic interaction of the 4f electrons with each other when there is no centrally symmetric contribution and H_{SO} is the spin-orbit coupling, respectively.

1.6 Suitable hosts for doping of rare earth ions

For efficient device applications, a lanthanide-based activator as a foreign ion and selection of a good host matrix are of vital importance. Energy level structures of most of the lanthanide ions are independent of the host matrices but phonon frequency plays an important role in nonradiative transitions via multi-phonon relaxation between closely spaced energy levels. The nonradiative relaxation rate can be evaluated by the energy gap law [6],

$$k_{nr} = \beta_{el} \exp(-\alpha(\Delta E - 2\hbar\omega_{max})) \quad (3)$$

where "β_{el}" and "α" are constants for a given host lattice, "ΔE" is the energy difference between the energy levels under consideration, and $\hbar\omega_{max}$ is the maximum phonon energy of the host. Hosts having low phonon energy decrease the possibility of nonradiative relaxation channel rate, which promotes to high luminescence efficiency. The phonon frequencies of some host matrices are listed in **Table 2**.

Host matrices	Phonon frequency (~cm^{-1})
Borate	1400
Phosphate	1100
Silicate	1000–1100
Germanate	800–975
$CaMoO_4$	800
Tellurite	600–850
YVO_4	600
Fluoride	500–600
$LaCl_3$	260
Bromide	175–190
$LaBr_3$	175
Iodide	160

Table 2.
Host matrices with maximum phonon frequency for different hosts [7].

2. Mechanism: down-conversion and up-conversion processes

2.1 Energy transfer in down conversion process

The downshifting process is a single photon process in which a high energy absorbed photon is converted into a low energy photon. This process is governed by the Stokes law. Luminescence of an optical material is ruled by the energy transfer process involved between acceptor and donor ions. Mostly, resonant radiative energy transfer, nonradiative energy transfer, and phonon assisted energy transfer between donor (D) and acceptor (A) are prominent energy transfer processes involved in the lanthanide ions. These important processes are represented in **Figure 2**.

2.2 Energy transfer in up-conversion

In 1960, Auzel has first observed the up-conversion (UC) phenomenon. It is a nonlinear anti-Stokes process in which two or more than two lower energy input pumped photons are converted into a high energy output photon [8]. There are some more different types of processes involved in the UC phenomenon. The more common energy transfer processes involved in the UC process are described as follows:

1. Ground state absorption (GSA)

   The RE ion lying in the ground state absorbs the excitation energy and gets promoted to a higher excited level. This process involves the single ion.

2. Excited state absorption (ESA)

   The atom in the excited level will be further promoted to the higher levels upon excitation. The ions from the higher level upon de-excitation emit an up-converted photon. This is also a single-ion process.

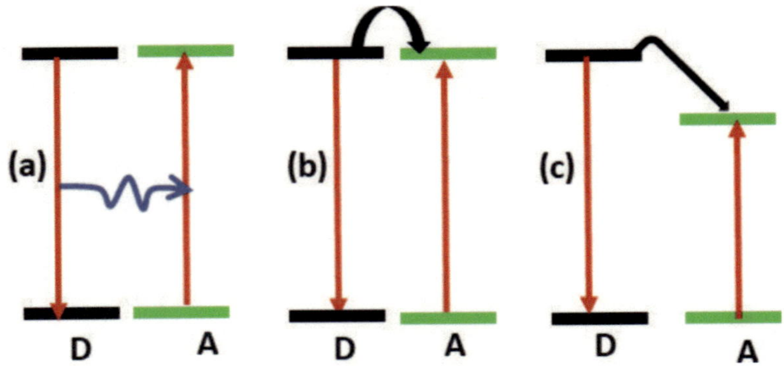

Figure 2.
(a) Resonant radiative energy transfer, (b) nonradiative energy transfer, and (c) phonon assisted energy transfer between donor (D) and acceptor (A) ions.

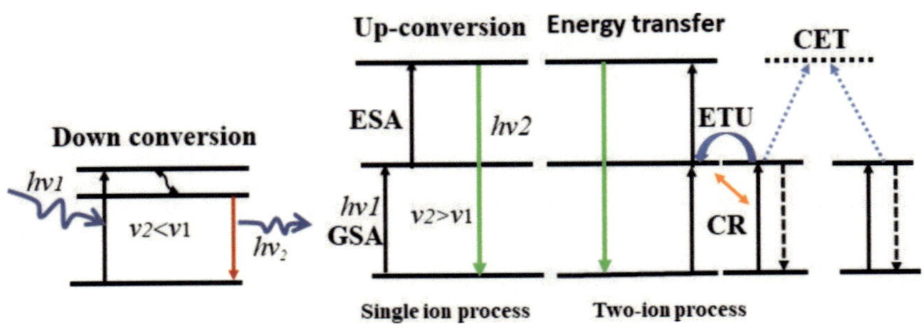

Figure 3.
Schematics of various processes, viz. down-conversion, up-conversion, and involved energy transfers.

3. Energy transfer up-conversion (ETU)

It is governed by participation of two different adjacent ions in which one ion acts as a sensitizer and absorbs the excitation energy, which is transferred to the neighboring ion acting as an activator (already lying in the excited state). The emitted photon can have higher energy excited photon.

4. Cross-relaxation process (CR)

If two sets of energy levels (in the same ion or in two different ions) are having the same energy difference between them, then there is certain probability of energy transfer and/or exchange occurring between the levels under consideration.

5. Co-operative energy transfer (CET)

When two active ions in a metastable intermediate level interact with each other and excite one ion into a high lying level and the other ion getting de-excited, in this process, very high concentration of doping is required.

The abovementioned energy transfer processes have been summarized in **Figure 3**.

3. Applications of the luminescent materials

3.1 Lanthanide doped nanophosphor materials as a solar spectral convertor for photovoltaic applications

There is tremendous research and advancement going on in the field of solar cell technology. It is still a challenging task for the photovoltaic (PV) community to prepare highly efficient and cost-effective conversion of solar energy into electricity. The spectral distribution of Sun light at Air Mass 1.5 global (AM 1.5G) constitutes with the photons having wide spectral range varying from ultraviolet to infrared (200–2500 nm) since PV cells are only sensitive to a small fraction of the incident solar photons. This is due to the fact that current solar materials respond to a narrow range of solar photons. Photons having energy higher than the bandgap of the material are only absorbed and excess energy is released in the form of heat [9].

In order to utilize the whole solar spectrum, a novel perspective through use of up and down conversion and quantum cutting has been exploited as an optical material that can act as a spectral converter since lanthanide-doped nanophosphor materials reduce the spectral mismatch losses, and therefore magnify the efficiency of photovoltaics.

In the up-conversion process, the sub-band gap of solar spectra is utilized. It is reported for Tm^{3+}, Er^{3+}, and Ho^{3+} co-doped with Yb^{3+} up-converters for wide band gap solar cells, while Er^{3+} doped up converters for c-Si solar cells. Moreover, Ln^{3+} doped up-convertors are under study for solar cells and have low conversion efficiency (~3%). The up-conversion efficiency may be magnified by employing plasmon resonance, quantum dots, and organic dyes [10–12].

Quantum cutting permits the effective use of the high energy photons which have above band gap value, which furnishes potential for efficiency enhancement of narrow band gap solar cells such as c-Si solar cells. An internal quantum efficiency of ~200% has been reported in lanthanide (Ln^{3+} = Tb, Er, Pr, Tm, Dy, Nd, and Ho) coupled with Yb^{3+} for narrow band c-Si solar cells [10].

Yb^{3+} ions act as an acceptor with NIR emission band at ~980 nm which has a band gap just above the band gap of c-Si. In order to improve the absorption cross-section which is common in lanthanide activators, other ions such as Ce^{3+}, Eu^{2+}, Yb^{3+}, and Bi^{3+} ions and/or hosts have been utilized for sensitization purpose [13–15]. The up-conversion-based material is kept in between a bifacial solar cell reflecting layer which is utilized to harvest the sub-bandgap of solar spectrum. The quantum cutting phenomenon is a linear process which is independent of the incident power. This permits the use of nonconcentrated sunlight [16]. Quantum-cutting-based optical materials are adjusted on the front surface of a mono-facial solar cell which permits the absorption of down converted photons by the solar cell.

Downshifting furnishes the accomplishment to enhance the spectral response of solar cells into short-wavelength regime. In order to enhance power-conversion efficiency in solar cells, RE-doped phosphors and glasses, QDs, and lanthanide-based organic complexes can be employed as solar concentrators and downshifting layers. Spectral conversion modules involving down-shifting (DS), quantum-cutting (QC), and up conversion (UC) materials for PV applications have been shown in **Figure 4** [10].

3.2 Quantum dots

In case of quantum dots (QDs), particle size is squeezed below its Bohr radius, which is termed as the quantum confinement effect. Under the quantum confinement effect, color of the QDs can be manipulated according to size and composition. It is due to the quantum confinement effect, emission color of QDs can be managed according to their size and composition [17]. Owing to these characteristics, QDs are applicable in solar cells, lasers, bio-imaging, and light-emitting diodes (LEDs) [18–21]. QDs can be mainly classified into Cd-based QDs as CdSe, CdTe and PbS, Cd-free InP, $CuInS_2$, and all inorganic $CsPbBr_3$ and mixed halide and $CH_3NH_3PbBr_3$ perovskite quantum dots (PQDs) [22–26]. PQDs comprise ABX_3 compositional formula unit in which A represents cesium (Cs), methyl ammonium

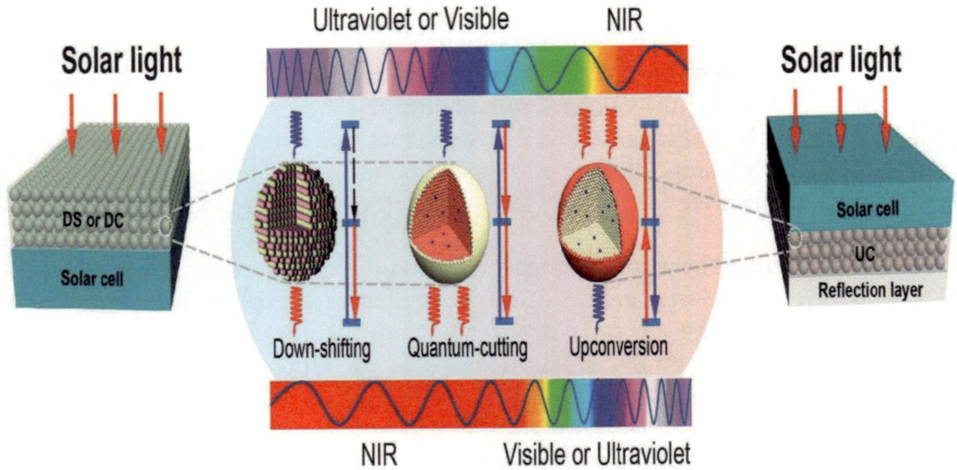

Figure 4.
Schematic spectral conversion module for photovoltaic utility involving down-shifting, quantum-cutting, and up-conversion-based luminescent materials. Reproduced with permission from reference [10].

(MA), while B represents Pb and/or Sn cations, and X represents Cl, Br, and/or I, that is, halide anions. QD-based LEDs with narrow FWHM can improve LED conversion efficiency from 5% to 15% and have correlated color temperature (CCT) values in the range of 5000–2700 K as compared to commercial red phosphor-based LEDs [27].

QLED devices gained curiosity in recent years over organic light-emitting diodes (OLEDs) due to their high color purity, relatively less cot, and lower energy consumption. The color gamut value for highly efficient QLEDs surpass 140% of the NTSC standard higher than that of commercial OLED. QD-based W-LEDs can reach ~104% enhancement of the NTSC standard [28]. High color purity of QLED devices annihilates the use of color filters, which magnify the power consumption value. QLED-based device performance is influenced by several constituents such as photoluminescence quantum yield (PLQY) of QD films, electron-hole pair efficiency from transport layer, and band arrangement of the device. A Cd-based red QLED device has been reported having ~20% external quantum efficiency (EQE) which is comparable to that of commercial OLEDs [29].

Instant development perspective for the first to fifth generations of white-light LED (W-LED) and lighting devices has been shown by Yoon et al. [30]. The very first generation of W-LED is mainly based on InGaN blue chip incorporated with yellow-emitting phosphors (YAG:Ce). Moreover, red color component of YAG-Ce-based W-LEDs is very feeble and it attains cool white light having a CRI value of ~70–80 [31, 32]. In order to enhance CRI values, in the 2nd generation, red emitting nitride phosphor-based LEDs are developed. Moreover, employed red nitride materials reveal lower efficiency since they possess a broader full width at half maximum (FWHM) and inherent subsidiary scattering loss. In the third generation, W-LEDs can be transformed as advanced in the form of backlight display. Red-QDs and narrow-band phosphor (KSF) are employed for the high-quality display applications. Prepared devices revealed wide color gamut as compared to traditional phosphor device. Red-QDs combined with YAG phosphor produce warm W-LEDs having a higher CRI value [33]. Wang et al. [34] incorporated red color CdS-ZnS: Cu QDs incorporated with YAG:Ce. In this study, the obtained device showed high values of CRI (Ra = 90) and CCT (4927 K). In order to replace Cd-based QDs, Cd-free $CuInS_2$/ZnS QDs incorporated with green Eu^{2+}:$BaSO_4$ phosphor have been prepared [35]. For fourth-generation development, the Illuminating Engineering Society of North America designed a new measurement module termed as TM-30-2015 having a defined color fidelity index (CFI, Rf) and color gamut index (CGI, Rg). In order to obtain an Rf value of over 90, it is difficult to produce warm-to-cool light W-LEDs. Warm-to-cool light W-LEDs should be produced for highly efficient white color. More recently, color-reproducible- and W-LEDs have been developed for next-generation lighting devices. The spectrum of QD-based W-LEDs should match daily fluctuation in sunlight under natural circadian rhythm for biological applications. The developed device is based on PQD having six-color white-light LEDs. In this module, six-colors, viz. cyan, green, yellowish green, amber, orange, and red of $CsPbX_3$ PQDs are used. The W-LED device thus packaged reveals high circadian with a tunable range and high color purity. The improved device shows a luminous efficiency (LE) of 58.8 lm w^{-1}, a CRI value of 95, CFI with Rf = 91.4, and CGI with Rg = 102 and a CCT value of 6459 K. Circadian luminescence is established on the day-night rhythmic change of natural-light environments [36].

3.3 Phosphor-based LEDs

LEDs are most commonly referred to as a p-n junction diode that can emit visible radiation under the suitable current application to the leads (electrical connection connecting a length of wire/metal pad). The mechanism is based on the phenomenon of electroluminescence. In the electroluminescence process, the material emits light in response of electric current. **Figure 5(a)** and **(b)** represent electronic symbolic representation and packaged conventional LEDs, respectively. LEDs can be divided into two broad categories namely inorganic and organic LEDs (OLEDs). The OLEDs can further divide into two main groups: (1) small molecule OLEDs and (2) polymer OLEDs. Phosphor-based W-LEDs are main attraction due to their excellent properties such as low power consumption, long life, compactness, and environmental friendliness. The successful implementation for W-LEDs could be done via composed blue InGaN LED chip and phosphors (yellow emitting $Y_3Al_5O_{12}:Ce^{3+}$ phosphor), and they are called phosphor converted W-LEDs [37].

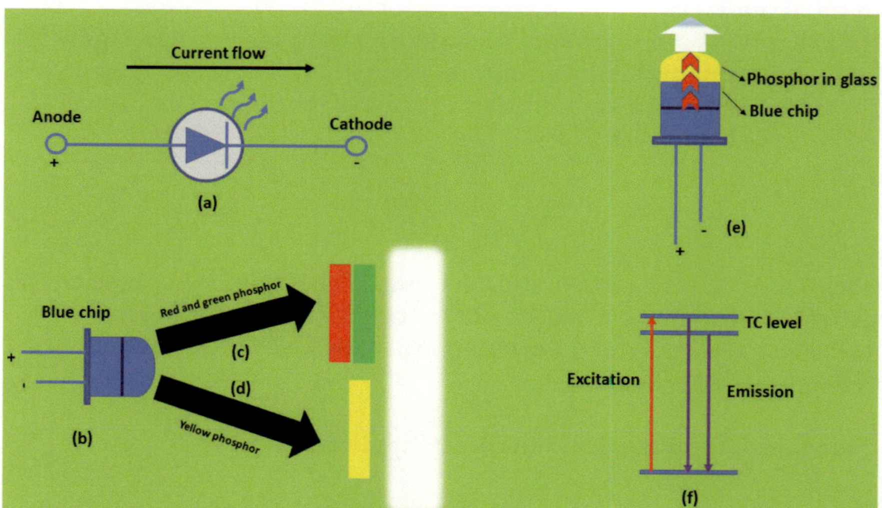

Figure 5.
(a) Electronic symbolic representation, (b) packaged conventional blue LED chip, (c) white light generation from red and green phosphor, (d) white light generation from yellow phosphor, (e) white light emission from PIG operated with blue chip, and (f) excitation and emission process originated from the two TC energy levels.

The approaches for the generation of white light is based on the combination of two or more visible colors (mainly primary color red, green, and blue) emitting phosphor. The advantage of uses of blue LED chip is such that it can be used as excitation light to phosphors as well as a source of blue component. **Figure 5(c)–(e)** shows the white light generation from red and green phosphor, white light generation from yellow phosphor, and white light emission from phosphor in glass (PIG) operated with blue chip. There are some necessities in the LED phosphors for devices packaging, viz. high efficiency by blue light excitation, stability in harsh environment, low thermal quenching, and negligible deviation in emission spectrum (peak position and line width). **Table 3** shows some synthesized phosphor applicable in the W-LED application. RE-doped optical glasses are noncrystalline amorphous solid luminescent materials that can be utilized in the production of lenses, prism, optical window, and other optical components. RE-incorporated glasses also exhibit good luminescence that can be utilized in LED fabrication. PIG is considered to be a reasonable substitute that can overcome the luminescent degradation and color aberration issues. PIG has been recognized as a good material for high-power W-LED products for long-term service.

3.4 Optical temperature sensing

The noncontact optical temperature sensing can be realized by using the RE ion-doped phosphor due the presence of thermally coupled (TC) energy levels. These TC levels have been used in the measurement of fluorescence intensity ratio (FIR) investigation. The Boltzmann's distribution population law can be applied to measure the FIR under the application of the external temperature. The following Boltzmann's distribution population law can be used for FIR measurement:

$$\mathrm{FIR} = \frac{I_u}{I_l} = \mathrm{B}\exp(-\Delta \mathrm{E}/\mathrm{kT}) \tag{4}$$

where "I_u" and "I_l" are the integrated fluorescence intensities of the emissions from upper and lower TC energy level transitions, respectively. "B" is the pre-exponential constant. "ΔE" is the energy gap of two TC energy levels. "k" and "T" are the Boltzmann's constant and absolute temperature, respectively. **Figure 5(f)** shows the excitation and emission process originated from the two TC energy levels. Sensor sensitivity (S) of an optical sensing material is an important parameter and can be expressed as follows:

$$\mathrm{S} = \frac{d\,(FIR)}{dT} \tag{5}$$

where all the terms have their usual meaning. The FIR technique is suitably applied in the RE ion-doped material for the detection of the temperature. The method is also applicable for the measurement of optical heating for laser-based hyperthermia application.

3.5 Quantum dots for medical and clinical application

The developments of the QDs have shown a promising impact on the biological application due to their highly efficient emissive properties. It is believed that the bare (uncoated) QDs cannot be applied directly in biological application due to their inherent toxicity. Due to their toxic behavior, their implication in clinical usages is restricted. The QDs have proven their utility in bioapplications and are a better alternative of the traditional fluorophores. QDs possess a very good

Materials	Dopant ions	References
$Y_3Al_5O_{12}$	Ce^{3+}	[38]
$Sr_2B_2O_5$	Tm^{3+}, Na^{+}	[39]
$K_3La(PO_4)_2$	Pr^{3+}	[40]
$Ba_3GdNa(PO_4)_3F$	Eu^{2+}	[41]
$ZnWO_4$	Eu^{3+}, Dy^{3+}	[42]
$Gd_4O_3F_6$	$Er^{3+}/Tm^{3+}/Yb^{3+}$	[43]
$LiNbO_3$	$Ho^{3+}/Yb^{3+}/Tm^{3+}$	[44]

Table 3.
Some synthesized phosphor applicable in the W-LED application.

photostability, tunable emission, and longer lifetime. Currently, the major scope is to synthesze more biocompatible QDs in such a way that they can be used in terms of size, bio-conjugation, surface functionality, and targeting facilities. The nanotechnology related to life-science and bioscience is called “Nano biotechnology”. The nanobiotechnology is a very fast grooving area and is a challenging task for the researchers working in this area. The remarkable result in the form of the useful medicine increases the attention in progress of novel therapy and diagnostic methods. The nanomaterials that preserve a broad scope in such novel areas are known as nanomedicines. Recently, the use of the QDs has been found to increase progressively in *in vitro* and *in vivo* based biomedical experiments. It is supposed that the QDs have specific properties such as adequate circulating lifetime and minimal nonspecific deposition and must retain their fluorescence for a satisfactorily long time.

The essentiality for the use of QDs in the surface passivated could be explained via three terms that are given as follows:

i. Surface-based properties (effect on the fluorescence and emission mechanism): the fluorescence property in the nanocrystalline particles can be quenched in bare QDs due to the surface defect caused by high surface energy associated with the nanoparticles [45, 46]. The main drawback comes due to the surface oxidation in bare QDs, photo-degradation, and leaching of metal ions via long-term contact of QDs to ionic media/cellular media causing metal ion toxicity effect [47]. To reduce this effect, it is essential to cap the surface of the QDs via a suitable covering agent (ZnS, etc.).

ii. Solubility: solubility is important for the effective optical property utilization of the QDs in the aqueous medium. The capping of the QDs provides an outer shell which improves the stability and reduces the solubility of the QDs in aqueous medium. The synthesis of QDs required a high temperature synthesize environment such as thermolysis, hot injection, and various other methods. Also, the size and agglomeration can be controlled via using the organic solvents (octane, hexane, trioctylphosphine oxide, octadecene, etc.) The stabilization can be achieved by using hydrophobic groups such as amines or phosphines. But the intrinsic solubility of these stabilized QDs is poor. To increase the solubility in the aqueous solution, coating of the QDs surface by hydrophilic ligand is important [48–50]. The hydrophilic ligand can be overcoated by using ligand exchange, surface silanization, and amphiphilic combination [50–52]. Medintz et al. have reported a detailed schematic of common QD solubilization and biofunctionalization [53].

Caps or ligand was used in biofunctionalization. Linkage to the QDs is given in pink color, biomolecule linking functionality is given in green color, and water solubility is given in blue color. Some models of the surface capping strategy and their interaction mechanism with the QDs and aqueous medium. For the cap exchange (top right) excess thiolated cap relocates the original TOP or TOPO coating via binding the ZnS surface layer with the thiol group and imparting hydrophilicity with the charged carboxyl (or other functionalities) which provides soluble colloidal QD dispersion.

iii. Targeted delivery of QDs: for targeted drug delivery (for in-vitro and in-vivo) of the QDs, the choice of use of the abovementioned processes depends on the application and also on the easy availability to carry out the strategy. It is observed that many alternative methods/strategies and ligands have been prepared for the specific biological applications. The essential steps required for the target specific delivery of the QDs are to reduce their effect over the nonrelevant cells, increase greater contrast in the localized area, and utilize more effective FRET-based processes to get better results [54].

The modifications of the QD surface for the labeling of a target can be done by using antibodies or enzymes, small molecules, and peptides [55, 56]. The versatile strategy has been found to be useful by antibodies. The requirements of the synthetic procedures can be utilized in the small molecules, and hence, it is complicated. It is observed that each process has its own advantage and disadvantage and thus no unique approach was seen to be ubiquitous for all applications [55–57]. The cell culture examination observed in most of the cases of vitro cell culture cannot show complete vivo tissue system. In this, various interactions between cell to cell and/or cell to matrix and many other diffusion /transport situations take place. The existence of the cells in the body exhibits 3D environments and is very serious for their growth and metabolism [58]. In the case of 2D cell culture on a flat substrate, all these interactions are effectively reduced, and thus, it confines their capacity to reiterate the appropriate level of in-vivo cellular retorts [59].

Lee et al. [60] first time performed a 3D spheroid culture-based NP toxicity testing arrangement by using 3D-liver-tissue-spheroid models with human hepatocarcinoma (HepG2) cells. The human liver is considered as the main organ for NP gathering. The detailed investigation of the toxicity of cadmium telluride (CdTe) QDs in cell culture and in a 3D model has been signified. The model can be supposed to be intermediates between in vitro and in vivo observations. In the case of 2D cell culture, the HepG2 cells are attached closely to each other which exist in normally flat shape after spreading on a well-plate. On CdTe QD exposer, morphology was found to be changed after 12 h culture. A sufficient number of cells are disturbed (contracted and curved/roosted) and some are detached from the well-plate. The similar observation was observed in the case of CTAB-AuNPs in 2D cell culture. But, the negligible change in the cellular structure was seen in the citrate-stabilized gold (Au) NPs. Also, no significant morphology was changed in the 3D spheroid culture. However, a small rough spheroid surface in the CdTe and CTAB-Au NP-treated cultures was found. For the clear view, the confocal microscopy was used to identify the live or dead cells. It was observed that the dead cells (red in color) in 2D culture transferred into granular shape and were missing from the plate after dropping their cell-cell and cell-substrate interactions. Nevertheless, the spheroid culture has not changed any different morphology, and apparently, a small number of dead cells were found in the surface of the spheroids in a rugged manner. The observation tells that in the spheroid culture small dead cells are present as compared to the 2D culture. This study suggests a comparative result between the 2D and 3D culture of HepG2 cells.

It is difficult to elucidate significant prediction by vivo effects based on the observation of vitro examination. The accurate mechanism involved in QD interaction with the body with metabolism distribution and final elimination after the uptake/delivery is still not cleared. Generally, NPs treated before the delivery follow long visit via altered body transport systems. Injection is considered as one of the main ways for QD-based vivo studies. The exogenous particle contains various serum proteins that can be carried out by blood. Protein corona formed by the interaction of the QD and protein can develop as a layer on the QD surface. Protein corona formed on the surface of the NPs affects biodistribution and biocompatibility. The available bloodstream helps in the delivery of NPs in various tissues and organs. The distribution is related to physio-anatomical structures of the vasculature and the physicochemical properties of the NPs. On completion of the distribution, the QDs undergo metabolic processing and will be evacuated from the body through the kidney and urine or feces. The excess sustained NPs inside of the body after holding long term periods can affect the normal function of organs/tissues. They can produce various toxicity levels such as metabolic toxicity, immunotoxicity, chronic organ toxicity, even genotoxicity, etc., [61]. The sufficient surface of vascular endothelial cells gives substantial space for the interactions with QDs specifically with negatively charged ones. The anionic QDs contain relatively low residence time in the blood and provide a huge accumulation in organs [62]. It has been observed that the high doses of the QDs produce contact activation and hence cause pulmonary vascular thrombosis by activating the coagulation cascade [63]. The importance of interaction of the QDs with blood cells is helpful in design for theranostic drug delivery resolutions. Fischer et al. have shown vivo mechanism by using two different (coated with mercaptoundecanoic acid and cross-linked with lysine and bovine serum albumin) coated QDs in rodents [64]. QDs coated with mercaptoundecanoic acid and cross-linked with lysine and bovine serum albumin was accumulated of about 40 and 99% in the liver after 90 min. A little amount of both the QDs was found in the spleen, kidney, and bone marrow. Moreover, the authors have not detected QDs in urine up to 10 days after intravenous dosing. Akerman et al. have observed a sufficient decrease of fluorescence via CdSe/ZnS QDs in animal studies using amphiphilic polymer and mercaptoacetic acid/peptide capped particles [65]. It has demonstrated slow metabolism degradation of QDs in mice. They observed a shift of the QD emission peak from red to blue upon long term exposure of over 2 years after injection of the QDs. In general, PEG-coated QDs decrease the toxicity in the cultured cells and provide good information related to lung tissue [66]. Tang et al. studied the biodistribution in mice and showed the effect of QD surface charge and their chemistry [67]. In spite of the surface chemistry and charge of the QDs, all the QDs produce toxic effect in liver, lung, and kidney after serious and long-term exposure. They observed that the PEG-coated QDs/negatively charged QDs accumulated in the liver. The positively charged QDs were located in the lung. The silica coted QDs after 5 days of injection show a very few of about 8.6% amount remaining in the hepatic tissue. But the QDs having approximate negligible protein corona were cleared through urine. The larger aggregated NPs can be transported to the liver and pass away via the bile excretion [68]. David Wegner and Hildebrandt have presented a broad review on different QD-based imaging applications for technological and the biological point of view [69]. Their review focused on the broad aspect of the QD utilization in in vivo and in vitro application extending from the super-resolution microscopy and single-particle tracking over in vitro cell and tissue imaging to in vivo examinations, drug or gene delivery to theranostic approaches and multimodal imaging.

Tsoi et al. have reported the blood clearance mechanism of managed hard nanomaterials or NPs relative to blood flow dynamics, organ microarchitecture, and cellular phenotype [70]. They observed that the velocity of the NPs which enters and traverse the liver reduces about 1000 times and shows 7.5 times NP interaction with the hepatic cells relative to peripheral cells. The three hard nanomaterial models, viz. quantum dots, gold nanoparticles, and silica nanoparticles, have been selected for the vivo examinations. **Figure 6** shows the distribution of QDs in the liver subsequent systemic intravascular injection. **Figure 6(a)** reflects the silver-stained section of a rat liver that was perfused 4 h post QDs injection. The accumulation of the QDs at a very higher amount was seen in the zone surrounding the portal triad than in the zone surrounding the central vein. Twenty-eight portal triad-central vein pairs have been analyzed. Here, one repeating unit of the liver microarchitecture is given for understanding **Figure 6(a)**. The blood flows can be seen from liver via the hepatic artery and portal vein that are positioned in the portal triad. The blood flows out of the liver through the central vein part of the liver. The zone of a radial distance of around 100 μm was taken for the tracing for each vascular unit of the liver. In **Figure 6(a)**, the scale bar is taken as 100 μm. The outline of the image processing that has been used to measure the accumulation of the QDs in the zone bordering of the portal triad and central vein has been shown in **Figure 6(b)**. Firstly, the zone surrounding each vascular structure has been removed by taking 100 μm radius from the vessel border. Secondly, the images were converted into a binary format and thresholder to isolate

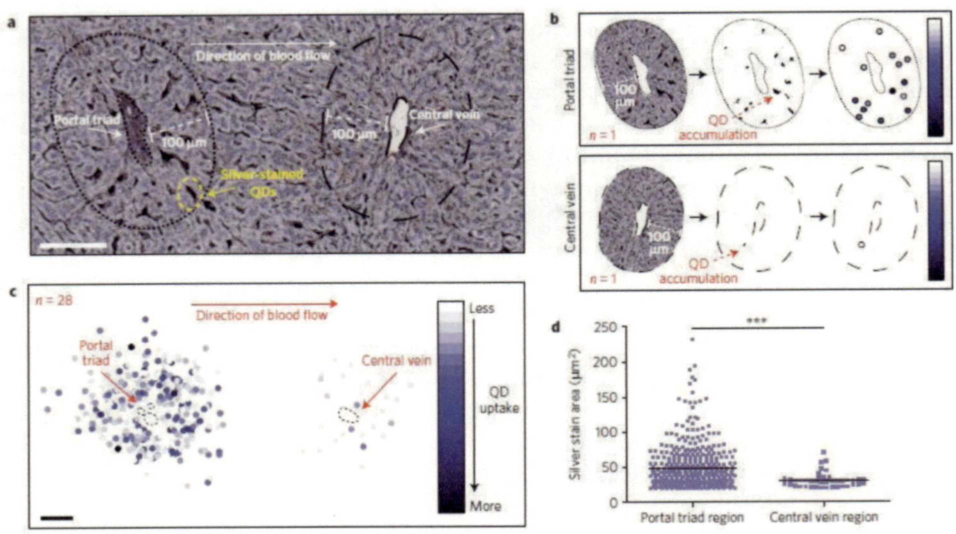

Figure 6.
Distribution of the QDs in liver succeeding systemic intravascular injection. (a) Silver-stained segment of a rat-liver that is perfused 4-h post-QD-injection (counterstained with hematoxylin). One repeating unit of the liver microarchitecture has been shown. Blood is flowing into the liver through the hepatic artery and portal vein, which are positioned in the portal triad. Blood is flowing out of the liver via the central vein. A zone with a radial distance of 100 μm was traced around each vascular unit. Scale bar is taken as 100 μm. (b) An outline of the image processing used to quantify QD growth in the zones nearby the portal triad and central vein. First, the zone neighboring of each vascular construction was removed using a radius of 100 μm from the vessel edge. Second, the image was transformed into a binary format and thresholder to isolate condensed silver. Finally, the area of each silver stain was controlled along with its coordinates (x, y) relative to the center of the vessel. The area of reduced silver corresponds to the amount of QDs gathering and is shown by a color range, where soft blue specifies a small amount of QD accumulation and dark blue shows a large amount of QDs gathering in each distinct position. (c) Twenty-eight portal triad-central vein pairs were analyzed and the results were combined. Scale bar is taken as 100 μm. (d) Scatter plot associated with the area of each silver staining in the zone surrounding the portal triad versus in the zone surrounding the central vein. Here, the plotted data are displayed same as shown in (c). Statistical significance was assessed by using a two-tailed unpaired t-test. Reproduced with permission from Ref. [70].

reduced silver. Then finally, the area of each silver strain has been scaled by using its coordinates (x, y) relative to the vessel center. The area of the QD accumulation has been given by a color spectrum in which the pale blue color represents a small amount of the QD accumulation and the dark blue reflects a large amount of QD accumulation in each individual location. The complete twenty-eight portal tried-central vein pairs were inspected. The combination of the results with the scale of the 100 μm is shown in the **Figure 6(c)**. The area of each region of silver marking in the zone surrounding the portal triad versus in the zone surrounding the central vein has been illustrated in **Figure 6(d)**. The data points taken are the same as shown in **Figure 6(c)**. Moreover, two-tailed unpaired t-test was considered for the statistical significance of the data. The mechanism of nanomaterial transport in the liver has been shown in **Figure 7**. The NPs that are injected in the bloodstream encounter the mononuclear phagocyte system (MRS), which is the group of organs with containing phagocytic cells. The degree of NP uptake within each MPS organ can be monitored from the intensity of the blue color (see **Figure 7**). When the NPs transit from the peripheral circulation to the liver, the observed velocity is found to decrease up to 1000 folds. This implies that the NPs could interact with the liver and hence their clearance is possible form the bloodstream. There exist a concentration gradient in the NPs along the length of the sinusoid, and the amount of the NPs leaving the liver via the central vein is lower than the amount that enters through the portal triad.

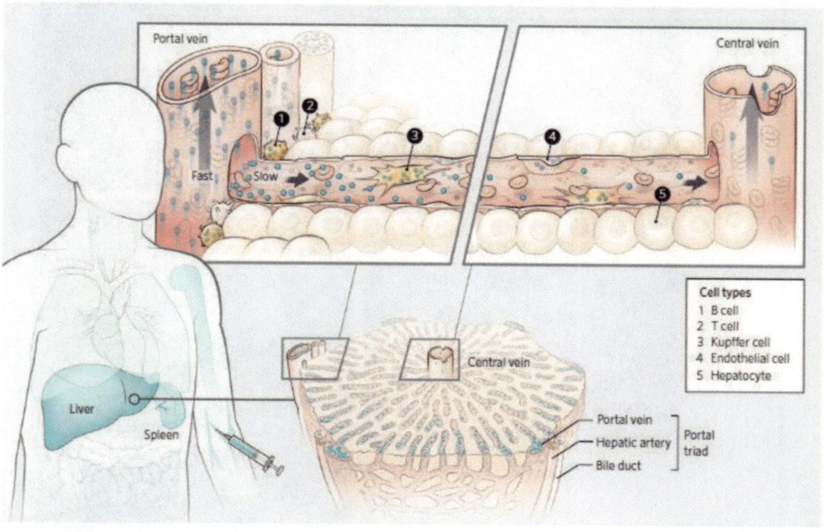

Figure 7.
Nanomaterial or NP mechanism for the transportation in the liver. The injected NPs into the bloodstream meet the MPS, which is a group of organs containing phagocytic cells. The emission intensity of the blue color in the figure imitates the mark of nanomaterial/NPs uptake within each MPS organ. As the NPs' transition from the peripheral circulation to the liver takes place, 1000 times their velocity is found to reduce. This allows the NPs to interact with a variety of cells, hence resulting in their gradual clearance from the blood stream. A concentration gradient of nanomaterials/NPs along the length of the sinusoid is seen, and the amount leaving the liver through the central vein is lower than that of the amount which enters via the portal triad (see the diagram of liver lobule, as shown in the bottom right). The cells, "B" and "T" are showing a border to the portal triad and are exposed to a high concentration of incoming NPs (can be seen from the schematic of a liver sinusoid, top). The difference in nanomaterial/NP uptake between these two cell types is because of the increased endocytic/phagocytic capacity of "B" cells as compared with the "T" cells. From the NPs, that escaped the first set of cellular interactions which is moving along the sinusoid and can come into contact with endothelial and Kupffer cells. Hepatocytes are separated from the bloodstream by a layer of fenestrated endothelial cells and do not seem to take up intact with the hard nanomaterials/NPs. Nanomaterials/NPs that have been escaped uptake during a pass through the liver return to the systemic circulation via the central vein and are ultimately carried back to the liver (or another MPS organ). This process is repeated until nanomaterial/NP clearance from the bloodstream is completed. Reproduced with permission from Ref. [70].

It can be seen that the "B" and "T" cells border the portal triad and open to a high concentration of incoming NPs. The observed differences that have been up-taken in both the NPs occur due to the increased endocytic or phagocytic capacity of "B" cells compared with "T" cells. The escaped NPs from the first set of cellular interaction can move along the sinusoid and hence can come into close contact with the endothelial and Kupffer cells. By a layer of fenestrated endothelial cells, the hepatocytes can be separated from the bloodstream and were not observed to take up hard NPs. The escaped NPs pass from the liver to the circulation process in the central vein and can return back to the liver or through the MPS organs. The steps involved in the process can be repeated by the NPs until the clearance from the blood stream is competed.

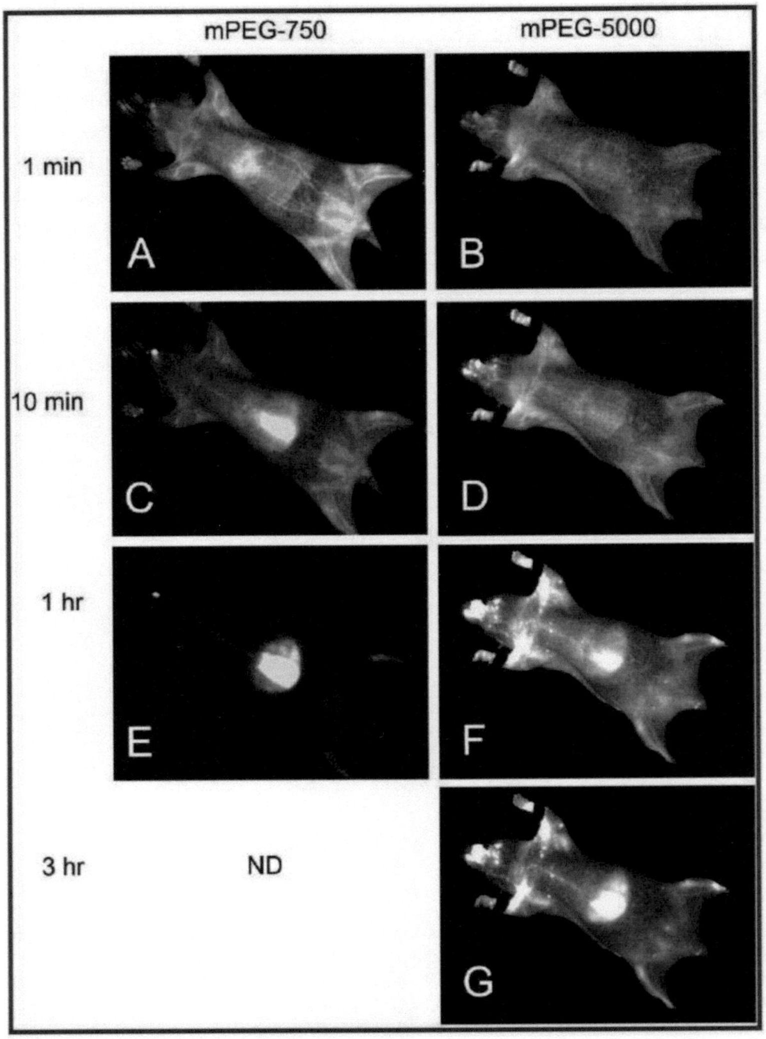

Figure 8.
Noninvasive imaging using 645 nm mPEG-750 QDs [left, (A, C, E)] and 655 nm-emitting mPEG-5000 QDs [right, (B, D, F, G)]. Nude mice have been taken to image at 1 min (A and B), 10 min (C and D), 1 h (E and F), and 3 h (G) post injection. It has been noted that even at 1 min, significant liver uptake is visible using mPEG-750 QDs (A), while even at 10 min there is little or no visible liver uptake using mPEG-5000 QDs (D). At 1 h, the difference between mPEG-750 QDs (E) and mPEG-5000 QDs is even more marked, with PEG-750 QDs completely cleared from the circulation, while mPEG-5000 QDs persist (F); even 3 h post injection, mPEG-5000 QDs remain in circulation (G). Since it is essentially found that all fluorescence was cleared from the circulation by 1 h, imaging of mPEG-750 QDs was terminated at 1 h after injection. In other experiments, we found that the pattern of deposition of mPEG-750 QDs can be seen at 1 h which remained stable for at least 24 h. Reproduced with permission from Ref. [71].

Ballou et al. have tested four different surface-coated QDs in in vivo imaging [71]. They have shown the QD localization using fluorescence imaging of living animals by necropsy, electron microscopy, frozen tissue sections (optical microscopy), and on scales oscillating from centimeters to nanometers for detection purpose by using only QDs. **Figure 8** shows the noninvasive imaging using 645 nm methoxy- or carboxy-terminated poly(ethylene glycol) amine, that is, mPEG-750 QDs (left, A, C, E) and 655 nm-emitting mPEG-5000 QDs (right, B, D, F, G). Nude mice were taken for the imaging at various time scales, 1 min, 10 min, 1 h, and 3 h, post injection for the study. It has been noted that even at 1 min, sufficient liver uptake is found to be visible by using mPEG-750 QDs (A); after 10 min only a very little or negligible visible liver uptake using mPEG-5000 QDs (D) was found. At 1 h time, an important difference between mPEG-750 QDs (E) and mPEG-5000 QDs is to be remarked, with PEG-750 QDs completely cleared from the circulation process, while mPEG-5000 QDs are to be seen persevered; even 3 h post injection, mPEG-5000 QDs remain in circulation. Moreover, it is crucial that all fluorescence is cleared from the circulation by up to 1 h; imaging of mPEG-750 QDs was found to complete at 1 h after injection. In other trials, they observed that the pattern of deposition of mPEG-750 QDs was realized at 1 h, which remained stable for at least 24 h.

The small QDs can penetrate in the blood-vessel walls and tissues and work as tracing peptides, in the in vivo antibody, tumor targeting, and as lymph node mapping due to their penetration of lymphatic walls [72–74]. But, the ability of the QDs to penetrate in other immunological barriers is not very much clear to date. Placental barrier (acts as a discriminating membrane to substances passing from maternal to fetal blood) has been considered an important process in humans/animals because of the prevention from the foreign materials via bacteria, viruses, and other organic or inorganic particles. Moreover, all foreign materials cannot effectively stop by this barrier. Some drugs cannot be employed in pregnant woman since it can be transported to the fetus across the placental barrier. At that place, they have the possibility to cause congenital malformations or miscarriage [75]. Chu et al. have reported [76] the inspection of intact QDs within the placental barrier. The transfer of QDs from pregnant animals to fetuses was reported successfully. The fetuses were cryosectioned at the time of observation. The ultrafine sections were distinguished with the support of fluorescence microscopy. Fluorescence was not at all observed in any tissues from QDs (red in color). The findings suggest that the QDs may be passed from female mice to their fetuses across the placental barrier. The small QDs have been transferred easily than their larger counterpart. The number of the QDs thus transferred will turn on the dosage. The increase in the QD dosage reveals enhancement in the transportation. Moreover, the QDs coated with the silica or other organic (such as polyethylene glycol PEG) layer could hinder the transportation process. This result shows that the clinical approach of QDs is limited in pregnant women.

4. Conclusion

The chapter includes a verity of luminescent materials that are useful in diverse applications such as development of the LEDs, display, temperature sensing, and biomedical applications. The spectroscopy of rare earth ions with their basic properties has been given in detail. The mechanism for understanding the emission processes involved in the RE activated optical materials has been discussed in a broad view by taking the concept of energy migration within the luminescent ions. The uses of these optical materials, viz. quantum dots and phosphors, have been incorporated. A broad view of the quantum dots has been accounted while considering the surface modification to show their biomedical application.

Acknowledgements

Abhishek Kumar Soni acknowledges the award of National Post-Doctoral Fellowship provided by the Science and Engineering Research Board (SERB), Government of India through N-PDF file No. PDF/2016/003419. Bheeshma Pratap Singh acknowledges the Department of Science and Technology (DST), New Delhi, India, through its DST-INSPIRE Faculty Award (IFA-17 MS-109) program.

Author details

Abhishek Kumar Soni* and Bheeshma Pratap Singh*
Chemistry Division, Bhabha Atomic Research Centre, Mumbai, India

*Address all correspondence to: aksoni.ism@gmail.com and bheeshmapratap@gmail.com

References

[1] Feldmann C, Justel T, Ronda CR, Schmidt PJ. Inorganic luminescent materials: 100 years of research and application. Advanced Functional Materials. 2003;**13**:511-516

[2] Blasse G, Grabmaier BC. Luminescent Materials. Berlin: Springer; 1994

[3] Ronda CR. In: Ronda CR, editor. Luminescence from Theory to Applications. Weinheim: Wiley-VCH Verlag GmbH & Co. KGaA; 2008. pp. 1-34

[4] Bhargava RN, Gallagher D, Hong X, Nurmikko A. Optical properties of manganese-doped nanocrystals of ZnS. Physical Review Letters. 1994;**72**:416

[5] Miniscalco WJ. Optical and Electronic Properties of Rare Earth Ions in Glasses. Optical Engineering. New York: Marcel Dekker Incorporated; 2001. pp. 17-112

[6] van Dijk JMF, Schuurmans MFH. On the nonradiative and radiative decay rates and a modified exponential energy gap law for *4f–4f* transitions in rare-earth ions. The Journal of Chemical Physics. 1983;**78**:5317

[7] Richards BS. Luminescent layers for enhanced silicon solar cell performance: Down-conversion. Solar Energy Materials & Solar Cells. 2006;**90**:1189

[8] Auzel F. Upconversion and anti-stokes processes with f and d ions in solids. Chemical Reviews. 2004;**104**:139-174

[9] Richards BS. Enhancing the performance of silicon solar cells via the application of passive luminescence conversion layers. Solar Energy Materials & Solar Cells. 2006;**90**:2329-2337

[10] Huang X, Han S, Huang W, Liu X. Enhancing solar cell efficiency: The search for luminescent materials as spectral converters. Chemical Society Reviews. 2013;**42**:173

[11] Boyer JC, van veggel FCJM. Absolute quantum yield measurements of colloidal $NaYF_4$: Er^{3+}, Yb^{3+} upconverting nanoparticles. Nanoscale. 2010;**2**:1417

[12] Zhang W, Ding F, Chou SY. Large enhancement of upconversion luminescence of $NaYF_4$:Yb^{3+}/Er^{3+} nanocrystal by 3D plasmonic nano-antennas. Advanced Materials. 2012;**24**:OP236

[13] Huang XY, Zhang QY. Near-infrared quantum cutting via cooperative energy transfer in Gd_2O_3:Bi^{3+},Yb^{3+} phosphors. Journal of Applied Physics. 2010;**107**:063505

[14] Rakov N, Maciel GS. Near-infrared quantum cutting in Ce^{3+}, Er^{3+}, and Yb^{3+} doped yttrium silicate powders prepared by combustion synthesis. Journal of Applied Physics. 2011;**110**:083519

[15] Zhou J, Zhuang Y, Ye S, Teng Y, Lin G, Zhu B, et al. Broadband downconversion based infrared quantum cutting by cooperative energy transfer from Eu^{2+} to Yb^{3+} in glasses. Applied Physics Letters. 2009;**95**:141101

[16] Fan B, Chlique C, Conanec OM, Zhang X, Fan X. Near-infrared quantum cutting material Er^{3+}/Yb^{3+} doped La_2O_2S with an external quantum yield higher than 100%. Journal of Physical Chemistry C. 2012;**116**:11652

[17] Butkus J, Vashishtha P, Chen K, Gallaher JK, Prasad SKK, Metin DZ, et al. The evolution of quantum confinement in $CsPbBr_3$ perovskite nanocrystals. Chemistry of Materials. 2017;**29**:3644

[18] Kim G-H, García de Arquer FP, Yoon YJ, Lan X, Liu M, Voznyy O, et al. High-efficiency colloidal quantum dot

photovoltaics via robust self-assembled monolayers. Nano Letters. 2015;**15**:7691

[19] Hoogland S, Sukhovatkin V, Howard I, Cauchi S, Levina L, Sargent EH. A solution-processed 1.53 mum quantum dot laser with temperature-invariant emission wavelength. Optics Express. 2006;**14**:3273

[20] Caruge JM, Halpert JE, Wood V, Bulovic V, Bawendi MG. Colloidal quantum-dot light-emitting diodes with metal-oxide charge transport layers. Nature Photonics. 2008;**2**:247

[21] Chen CW, Wu DY, Chan YC, Lin CC, Chung PH, Hsiao M, et al. Evaluations of the chemical stability and cytotoxicity of $CuInS_2$ and $CuInS_2$/ZnS core/shell quantum dots. Journal of Physical Chemistry C. 2015;**119**:2852

[22] Nan W, Niu Y, Qin H, Cui F, Yang Y, Lai R, et al. Crystal structure control of zinc-blende CdSe/CdS core/shell nanocrystals: Synthesis and structure-dependent optical properties. Journal of the American Chemical Society. 2012;**134**:19685

[23] Wu T, He K, Zhan Q , Ang S, Ying J, Zhang S, et al. MPA-capped CdTe quantum dots exposure causes neurotoxic effects in nematode *Caenorhabditis elegans* by affecting the transporters and receptors of glutamate, serotonin and dopamine at the genetic level, or by increasing ROS, or both. Nanoscale. 2015;**7**:20460

[24] Moreels I, Justo Y, De Geyter B, Haustraete K, Martins JC, Hens Z. Size-tunable, bright, and stable PbS quantum dots: A surface chemistry study. ACS Nano. 2011;**5**:2004

[25] Schmidt LC, Pertegás A, GonzálezCarrero S, Malinkiewicz O, Agouram S, Mínguez Espallargas G, et al. Nontemplate synthesis of $CH_3NH_3PbBr_3$ perovskite nanoparticles. Journal of the American Chemical Society. 2014;**136**:850

[26] Protesescu L, Yakunin S, Bodnarchuk MI, Krieg F, Caputo R, Hendon CH, et al. Nanocrystals of cesium lead halide perovskites ($CsPbX_3$, X = Cl, Br, and I): Novel optoelectronic materials showing bright emission with wide color gamut. Nano Letters. 2015;**15**:3692

[27] Shimizu KT, Böhmer M, Estrada D, Gangwal S, Grabowski S, Bechtel H, et al. Toward commercial realization of quantum dot based white light-emitting diodes for general illumination. Photonics Research. 2017;**5**:A1

[28] Jang E, Jun S, Jang H, Lim J, Kim B, Kim Y. White-light-emitting diodes with quantum dot color converters for display backlights. Advanced Materials. 2010;**22**:3076

[29] Dai X, Zhang Z, Jin Y, Niu Y, Cao H, Liang X, et al. Solution-processed, high-performance light-emitting diodes based on quantum dots. Nature. 2014;**515**:96

[30] Yoon HC, Oh JH, Lee S, Park JB, Do YR. Circadian-tunable perovskite quantum dot-based down-converted multi-package white LED with a color fidelity index over 90. Scientific Reports. 2017;**7**:2808

[31] Lin CC, Liu RS. Advances in phosphors for light-emitting diodes. Journal of Physical Chemistry Letters. 2011;**2**:1268

[32] Wang B, Lin H, Xu J, Chen H, Wang Y. $CaMg_2Al_{16}O_{27}:Mn^{4+}$-based red phosphor: A potential color converter for high-powered warm W-LED. ACS Applied Materials & Interfaces. 2014;**6**:22905

[33] Zhu H, Lin CC, Luo W, Shu S, Liu Z, Liu Y, et al. Highly efficient non-rare-earth red emitting phosphor for warm white light-emitting diodes. Nature Communications. 2014;**5**:4312

[34] Wang X, Yan X, Li W, Sun K. Doped quantum dots for white-light-emitting diodes without reabsorption of multiphase phosphors. Advanced Materials. 2012;**24**:2742

[35] Chuang PH, Lin CC, Liu R-S. Emission-tunable $CuInS_2$/ZnS quantum dots: Structure, optical properties, and application in white light-emitting diodes with high color rendering index. ACS Applied Materials & Interfaces. 2014;**6**:15379

[36] Wang HC, Bao Z, Tsai HY, Tang AC, Liu RS. Perovskite quantum dots and their application in light-emitting diodes. Small. 2017;**14**:1702433

[37] Soni AK, Mahata MK. Photoluminescence and cathodoluminescence studies of Er^{3+}-activated strontium molybdate for solid-state lighting and display applications. Materials Research Express. 2017;**4**:126201

[38] Niu X, Xun J, Prog ZY. The spectroscopic properties of $Dy^{3+}$and Eu^{3+} co-doped $Y_3Al_5O_{12}$ (YAG) phosphors for white LED. Natural Science. 2015;**25**:209

[39] Cai L, Ying L, Zheng J, Fan B, Chen R, Chen C. Luminescent properties of $Sr_2B_2O_5$: Tm^{3+}, Na^+ blue phosphor. Ceramics International. 2014;**40**:6913

[40] Bedyal AK, Ramteke DD, Kumar V, Swart HC. Blue photons excited highly chromatic red light emitting $K_3La(PO_4)_2$:Pr^{3+} phosphors for white light emitting diodes. Materials Research Bulletin. 2018;**103**:173

[41] Li C, Dai J, Huang J, Deng D, Yu H, Wang L, et al. Crystal structure, luminescent properties and white light emitting diode application of $Ba_3GdNa(PO_4)_3F$:Eu^{2+} single-phase white light-emitting phosphor. Ceramics International. 2016;**42**:6891

[42] Zhai Y, Wang M, Zhao Q, Yu J, Li X. Fabrication and luminescent properties of $ZnWO_4$:Eu^{3+}, Dy^{3+} white light-emitting phosphors. Journal of Luminescence. 2016;**172**:161

[43] Passuello T, Piccinelli F, Pedroni M, Bettinelli M, Mangiarini M, Naccache R, et al. White light upconversion of nanocrystalline Er/Tm/Yb doped tetragonal $Gd_4O_3F_6$. Optical Materials. 2011;**33**:643

[44] Xing L, Wang R, Xu W, Qian Y, Xu Y, Yang C, et al. Upconversion white-light emission in Ho^{3+}/Yb^{3+}/Tm^{3+} codoped $LiNbO_3$ polycrystals. Journal of Luminescence. 2012;**132**:1568

[45] Manna L, Scher EC, Li LS, Alivisatos AP. Epitaxial growth and photochemical annealing of graded CdS/ZnS shells on colloidal CdSe nanorods. Journal of the American Chemical Society. 2002;**124**:7136

[46] Hess BC, Okhrimenko IG, Davis RC, Stevens BC, Schulzke QA, Wright KC. Surface transformation and photoinduced recovery in CdSe nanocrystals. Physical Review Letters. 2001;**86**:3132

[47] Singh S, Sharma A, Robertson GP. Realizing the clinical potential of cancer nanotechnology by minimizing toxicologic and targeted delivery concerns. Cancer Research. 2012;**72**:5663

[48] Gao X, Cui Y, Levenson RM, Chung LW, Nie S. In vivo cancer targeting and imaging with semiconductor quantum dots. Nature Biotechnology. 2004;**22**:969

[49] Kairdolf BA, Mancini MC, Smith AM, Nie S. Minimizing nonspecific cellular binding of quantum dots with hydroxyl-derivatized surface coatings. Analytical Chemistry. 2008;**80**:3029

[50] Thanha NTK, Green LAW. Functionalisation of nanoparticles for

biomedical applications. Nano Today. 2010;**5**:213

[51] Li H, Shih WY, Shih WH. Highly photoluminescent and stable aqueous ZnS quantum dots. Industrial and Engineering Chemistry Research. 2010;**49**:579

[52] Lukehart CM, Scott RA, editors. Nanomaterial Inorganic and Bioinorganic Perspective. John Wiley & Sons; 2013. pp. 512-513. Available from: https://www.wiley.com/en-us/Nanomaterials%3A+Inorganic+and+Bioinorganic+Perspectives-p-9780470516447

[53] Medintz IL, Uyeda HT, Goldman ER, Mattoussi H. Quantum dot bioconjugates for imaging, labelling and sensing, Nature Materials. 2005;**4**:435

[54] Jamieson T, Bakhshi R, Petrova D, Pocock R, Imani M, Seifalian AM. Biological applications of quantum dots. Biomaterials. 2007;**28**:4717

[55] Rosenthal SJ, Chang JC, Kovtun O, McBride JR, Tomlinson ID. Biocompatible quantum dots for biological applications. Chemistry & Biology. 2011;**18**:10-24

[56] Goldman ER, Anderson GP, Tran PT, Mattoussi H, Charles PT, Mauro JM. Conjugation of luminescent quantum dots with antibodies using an engineered adaptor protein to provide new reagents for fluoroimmunoassays. Analytical Chemistry. 2002;**74**:841

[57] Chana WCW, Maxwell DJ, Gao X, Bailey RE, Han M, Nie S. Luminescent quantum dots for multiplexed biological detection and imaging. Current Opinion in Biotechnology. 2002;**13**:40

[58] Gartner PL, Hiatt LJ. Color Text Book of Histology. Philadelphia, PA: Saunder; 2001

[59] Lee J, Cuddihy MJ, Kotov NA. Three-dimensional cell culture matrices: State of the art. Tissue Engineering Part B. 2008;**14**:61

[60] Lee J, Lilly GD, Doty RC, Podsiadlo P, Kotov NA. In vitro toxicity testing of nanoparticles in 3D cell culture. Small. 2009;**5**:1213

[61] Li J, Chang X, Chen X, Gu Z, Zhao F, Chai Z, et al. Toxicity of inorganic nanomaterials in biomedical imaging, Biotechnology Advances. 2014;**32**:727

[62] Praetner M, Rehberg M, Bihari P, Lerchenberger M, Uhl B, Holzer M, et al. The contribution of the capillary endothelium to blood clearance and tissue deposition of anionic quantum dots in vivo. Biomaterials. 2010;**31**:6692

[63] Geys J, Nemmar A, Verbeken E, Smolders E, Ratoi M, Hoylaerts MF, et al. Acute toxicity and prothrombotic effects of quantum dots: Impact of surface charge. Environmental Health Perspectives. 2008;**116**:1607

[64] Fischer HC, Liu L, Pang KS, Chan WCW. Pharmacokinetics of nanoscale quantum dots: In vivo distribution, sequestration, and clearance in the rat. Advanced Functional Materials. 2006;**16**:1299

[65] Akerman ME, Chan WCW, Laakkonen P, Bhatia SN, Ruoslahti E. Proceedings of the National Academy of Sciences of the United States of America. 2002;**99**:12617

[66] Ho C-C, Chang H, Tsai H-T, Tsai M-H, Yang C-S, Ling Y-C, et al. Quantum dot 705, a cadmium-based nanoparticle, induces persistent inflammation and granuloma formation in the mouse lung. Nanotoxicology. 2013;**7**:105

[67] Tang Y, Han S, Liu H, Chen X, Huang L, Li X, et al. The role of surface chemistry in determining in vivo biodistribution and toxicity of CdSe/ZnS core-shell quantum dots. Biomaterials. 2013;**34**:8741

[68] Chen Z, Chen H, Meng H, Xing G, Gao X, Sun B, et al. Bio-distribution and metabolic paths of silica coated CdSeS quantum dots. Toxicology and Applied Pharmacology. 2008;**230**:364

[69] David Wegner K, Hildebrandt N. Quantum dots: bright and versatile in vitro and in vivo fluorescence imaging biosensors. Chemical Society Reviews. 2015;**44**:4792

[70] Tsoi KM, MacParland SA, Ma XZ, Spetzler VN, Echeverri J, Ouyang B, et al. Mechanism of hard-nanomaterial clearance by the liver. Nature Materials. 2016;**15**:1212

[71] Ballou B, Lagerholm BC, Ernst LA, Bruchez MP, Waggoner AS. Noninvasive imaging of quantum dots in mice. Bioconjugate Chemistry. 2004;**15**:79

[72] Cai W, Shin D-W, Chen K, Gheysens O, Cao Q, Wang SX, et al. Peptide-labeled near-infrared quantum dots for imaging tumor vasculature in living subjects. Nano Letters. 2006;**6**:669

[73] Kerman M, Chan WCW, Laakkonen P, et al. Nanocrystal targeting in vivo. PNAS. 2002;**99**:12617

[74] Kim S, Lim YT, Soltesz EG, Grand AMD, Lee J, Nakayama A, et al. Near-infrared fluorescent type II quantum dots for sentinel lymph node mapping. Nature Biotechnology. 2004;**22**:93

[75] Briggs GG, Freeman RK, Yaffe SJ. Drugs in Pregnancy and Lactation: A Reference Guide to Fetal and Neonatal Risk. Lippincott Williams & Wilkins; 2012. Available from: https://books.google.co.in/books?hl=hi&lr=&id=ScPvM03B3lUC&oi=fnd&pg=PA181&dq=Briggs+GG,+Freeman+RK,+Yaffe+SJ.++Drugs+in+Pregnancy+and+Lactation.+Lippincott+Williams+%26+Wilkins%3B+2001&ots=tJT5pDgZrj&sig=xocU0n9TwTWdK6WLCuOVnxz0tjc#v=onepage&q&f=false

[76] Chu M, Wu Q, Yang H, Yuan R, Hou S, Yang Y, et al. Transfer of quantum dots from pregnant mice to pups across the placental barrier. Small. 2010;**6**:670

Chapter 2

Triboluminescence: Materials, Properties, and Applications

Zhaofeng Wang and Fu Wang

Abstract

Triboluminescence is one of the types of luminescence that could be activated by mechanical stress. Considering the rising research efforts and achievements in recent years, this chapter provides an overview on the study of triboluminescence. The first part gives a background description regarding the history, research status, and advantages of triboluminescence. Then, we summarize the material systems for triboluminescence in both organics and inorganics. In the third part, we review the properties of triboluminescence, particularly on the unique characteristics and their improvements. Finally, we give a comprehensive summary on the developments of triboluminescent devices for applications in various fields in terms of mechanical engineering, energy, biological monitoring, and sensors as well as lighting, imaging, and displaying.

Keywords: triboluminescence, crystals, spectral characteristics, cycling stability, advanced applications

1. Introduction

Triboluminescence (TL) refers to the phenomenon that materials could emit light when they are mechanically stimulated, such as rubbing, grinding, impact, stretching, and compression [1–3]. TL was first recorded by Francis Bacon in 1605 when breaking the sugar crystals [4]. After that, TL has been found in many solids, such as rocks, quartz, alkaline halide, molecular crystals, and some organic materials [5]. It is estimated that nearly 50% of inorganic compounds and 30% of organic molecular solids have been confirmed to have TL [6]. Because TL could be directly activated by the widely existed mechanical activities in daily life without requiring artificial optical/electrical sources, TL shows great advantages in energy saving and environmental protection [7].

In general, TL could be classified into three types, i.e., fracture TL, plastic TL, and elastic TL [8], as illustrated in **Figure 1**. Among them, the elastic TL has gained the most attention because of its structure nondestructive characteristic which is crucial for practical applications. The present researches of TL are mainly focused on the development of novel TL materials and the performance improvement in terms of brightness, color manipulation, and cyclic stability [9–11]. Based on the efforts in the above aspects, a variety of decent applications of TL materials have been achieved in recent years, covering the fields of mechanical engineering, energy, biological monitoring, and sensors as well as lighting, imaging, and displaying.

In this chapter, we provide an overview of TL, regarding the materials, properties, and applications. Since TL covers a large range from organics to inorganics with emitting types from fracture TL to plastic TL and elastic TL, most of the

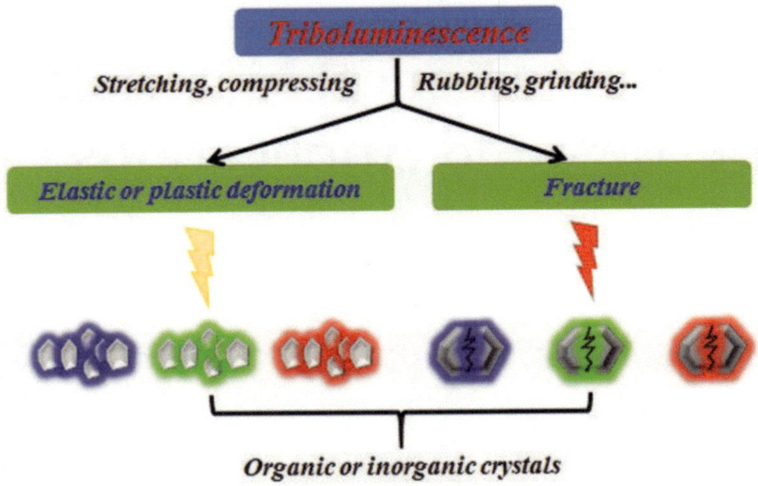

Figure 1.
Illustration of the fracture, plastic, and elastic deformation-induced TL in organic and inorganic crystals.

content of the chapter was focused on the elastic TL of inorganic solids in which the most significant progresses have been made during recent years. We hope that this chapter could provide a deep understanding of TL and stimulated new ideas for further researches.

2. TL materials

2.1 Organic crystals and organometallic compounds

Organic crystals and organometallic compounds represent an important part of TL materials. About 19% of organics and 37% of aromatic compounds are estimated to have TL [12]. According to molecular structure, the TL organic crystals could be divided into nonaromatic organic crystals and aromatic compounds. The main nonaromatic organics include sugar (e.g., D-glucose, lactose, maltose, L-rhamnose, sucrose), tartaric acid/tartrate (e.g., ammonium tartrate, sodium tartrate) and other nonaromatic organics (e.g., L-ascorbic acid, cholesteryl salicylate, cholestenol, ammonium oxalate, disodium hydrogen citrate, aniline hydrochloride) [13–15]. The main aromatic compounds are coumarin, acenaphthene, phthalic anhydride, phenanthrene, phenol derivatives, 9-anthryl carbinol, N-phenyl-substituted imides, carbazole derivatives, hexaphenylcarbodiphosphorane $(Ph_3P)_2C$, and some aggregation-induced emission compounds (e.g., tetraphenylethene compounds, N-substituted phenothiazine, aryl dioxaborolane, N-substituted dihydroacridine) [16–18]. The above aromatic compounds always possess distinctive TL characteristics because of their peculiar molecular structure, and their TL should arise from the spin-allowed/spin-forbidden electron transition of molecular excited state (π-π^* transition), likewise with their photoluminescence (PL). Moreover, impurities play special roles in TL of some compounds.

Organometallic compounds, including rare earth and transition metal complexes, have also featured TL. The typical examples are some β-diketone complexes of Ln^{III} ion (Ln = Eu, Sm, Pr, Yb, Tb, Gd, or Nb). Among them, the europium complexes (EuD_4TEA and its doped forms) generate extremely bright and daylight-visible red-orange TL, which is much stronger than that of the others [19]. But these complexes show a very sharp emission band corresponding to the f-f transition of Eu^{3+} ions. The transition metal-based complexes are mainly Mn^{II}, $copper^{I}$, and Pt^{II} complexes, such as $Mn(Ph_3PO)_2X_2$ (X = Cl, Br), $(MePh_3P)_2MnCl_4$, Cu(NCS)$(py)_2(PPh_3)$, and Pt(ipyim)(bipz), which give a broad emission band [20].

2.2 Inorganic compounds

The inorganic TL compounds are composed by hosts and doping luminescent centers. The inorganic hosts include the halides (e.g., KCl, KBr, NaF, RbBr, and RbI [21]), oxides (e.g., Al_2O_3 [22] and ZrO_2 [23]), sulfides (e.g., ZnS [24]), oxysulfides (e.g., CaZnOS [10] and BaZnOS [25]), aluminates (e.g., $SrAl_2O_4$ [1], $Sr_3Al_2O_6$ [7], and $CaYAl_3O_7$ [9]), silicates (e.g., $Sr_2MgSi_2O_7$ and $SrCaMgSi_2O_7$ [26]), phosphates (e.g., Li_3PO_4 [27] and $SrMg_2(PO_4)_2$ [28]), borates (e.g., BaB_4O_7 [29]), titanates (e.g., $BaTiO_3$ and $CaTiO_3$ [30]), niobates (e.g., $Ca_2Nb_2O_7$ [31] and $LiNbO_3$ [32]), stannates (e.g., Sr_2SnO_4 [33]), sulfates (e.g., $BaSO_4$ [34]), and oxynitrides (e.g., $BaSi_2O_2N_2$ [35]). Rare earth ions are the common doped ions in inorganic TL compounds, such as Eu^{2+}, Eu^{3+}, Pr^{3+}, Dy^{3+}, Ce^{3+}, Tb^{3+}, Er^{3+}, and Sm^{3+} [4]. The other metal ions, like Mn^{2+}, Cu^{+}, and Ti^{4+} ions [23, 36], are also employed as the luminescent centers in inorganic TL compounds. To date, the well-recognized inorganic compounds with bright TL are $SrAl_2O_4:Eu^{2+}$, Dy^{3+} (SAOED), and $ZnS:Mn^{2+}/Cu^{+}$.

3. TL properties

3.1 Spectral characteristics

In many organic and inorganic systems, the TL spectra are consistent with the PL spectra, suggesting they possess the same emitting processes. The differences between TL and PL lie in the excitation/activation processes that TL originates from the release of the trapped carries or the piezoelectric effect under mechanical stimuli. In some systems, like $BaZnOS:Mn^{2+}$ [25], the compression-induced TL and rubbing-induced TL exhibit 24 nm and 48 nm blueshift, respectively, compared to that of PL (**Figure 2**). Such phenomenon could be ascribed to the conduction band and valence band tailoring by piezoelectric fields.

In piezoelectric materials, there is also obvious difference on the concentration quenching between PL and TL. For example, the quenching concentrations of Pr^{3+} in $CaNb_2O_6$, $Ca_2Nb_2O_7$, and $Ca_3Nb_2O_8$ for TL are 0.25 mol%, 0.1 mol%, and 0.075 mol%, respectively, while the values for PL are 0.5 mol%, 0.3 mol%, and 0.1 mol%, respectively [31]. The decreased quenching concentration of TL was attributed to the participation of piezoelectric field in delivering the energy from traps to quenching centers.

3.2 Cycling stability

The TL of organic molecules or complexes mostly originates from the fracture of crystals, and thus there is no cycling stability for such materials. For the TL along with the nondestructive structure, mainly referring to the piezoelectric effect and de-trapping-induced TL, the cycling stability is particularly important. The TL aroused by piezoelectric effects usually exhibits stable luminescence when activated by cyclic mechanical tests [37]. For example, the Pr^{3+}-doped $LiNbO_3$ could keep its TL intensity for more than 100 cycles [32]. ZnS:Cu/PDMS composites could maintain the TL intensity up to 30,000 cycles of stretching, and the intensity still reached 65% of the initial one without a color change even after 100,000 cycles of

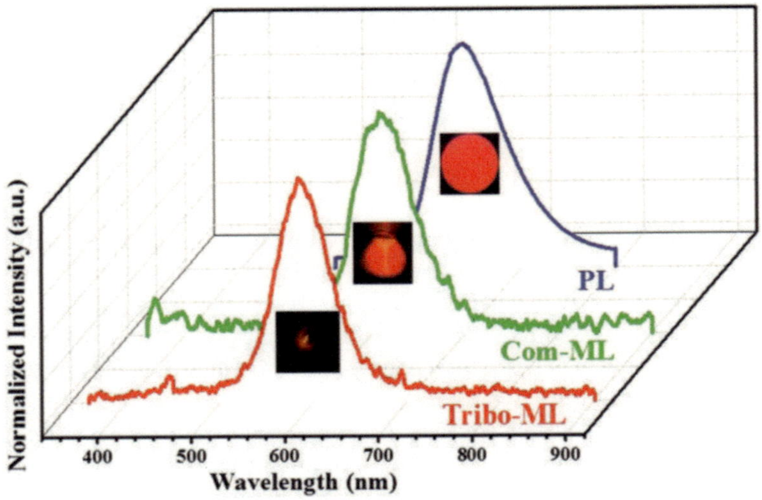

Figure 2.
Spectral comparison of the PL, compression-induced TL, and rubbing-induced TL in BaZnOS:Mn2⁺. Reproduced by permission of the Royal Society of Chemistry [25].

tests [38]. However, for the TL aroused by the de-trapping of carriers in structure, intensity degradation would be serious during cycling tests, i.e., such materials showed poor cycling stability [39, 40]. To overcome the above issue, great efforts have been made based on the TL mechanism in terms of the de-trapping processes. Researchers proposed a strategy to improve the cycling stability of the de-trapping-induced TL by applying an extra UV irradiation source to ensure the balance between the trapping and de-trapping of carries [11]. The power density played a key role to stabilize the TL intensity, and the effective power density was determined to be 1000 mW/cm^2 as shown in **Figure 3** [41].

3.3 Intrinsic structure-dependent TL

The TL characteristics could be directly modulated by varying the concentration of luminescent centers. Generally, there is a concentration quenching phenomenon in terms of TL intensity. In $CaZnOS:Mn^{2+}$, the increase of the doping concentration of Mn^{2+} could not only vary the TL intensity with a trend that increases first and then decreases but also arouse a redshift on the TL spectra with the emitting color manipulated from orange to red [10]. In addition, the chemical composition of the hosts, namely, the variation of the defect phases or traps, could also cause significant variations on the TL intensity and color. In $(Ba,Ca)TiO_3:Pr^{3+}$, the co-dopant of trivalent rare earth ions, such as La^{3+}, Y^{3+}, Nd^{3+}, Gd^{3+}, Yb^{3+}, and Lu^{3+}, could greatly improve the TL intensity, in which Gd^{3+} could enhance the intensity more than 61% [30]. This is because that the co-dopant of the above ions could increase the concentration of the carries in traps and thus lead to more luminescence emitted under mechanical stimuli. In $Sr_2MgSi_2O_7:Eu^{2+}$, when part of Sr^{2+} was substituted by Ca^{2+} or Ba^{2+}, the TL intensity and emitting color could be adjusted simultaneously [26]. $SrBaMgSi_2O_7:Eu^{2+}$ showed the lowest TL intensity compared to that of $Sr_2MgSi_2O_7:Eu^{2+}$ and $SrCaMgSi_2O_7:Eu^{2+}$. The replacement of Sr^{2+} by Ca^{2+} or Ba^{2+} in $Sr_2MgSi_2O_7:Eu^{2+}$ could further manipulate the emission band in a wide range from 440 nm to 499 nm. Researches also showed that the TL performance is dependent on the crystal size. In sucrose crystals, the TL intensity significantly increased with increasing crystal size (**Figure 4**), which could be explained by piezoelectric mechanism [42].

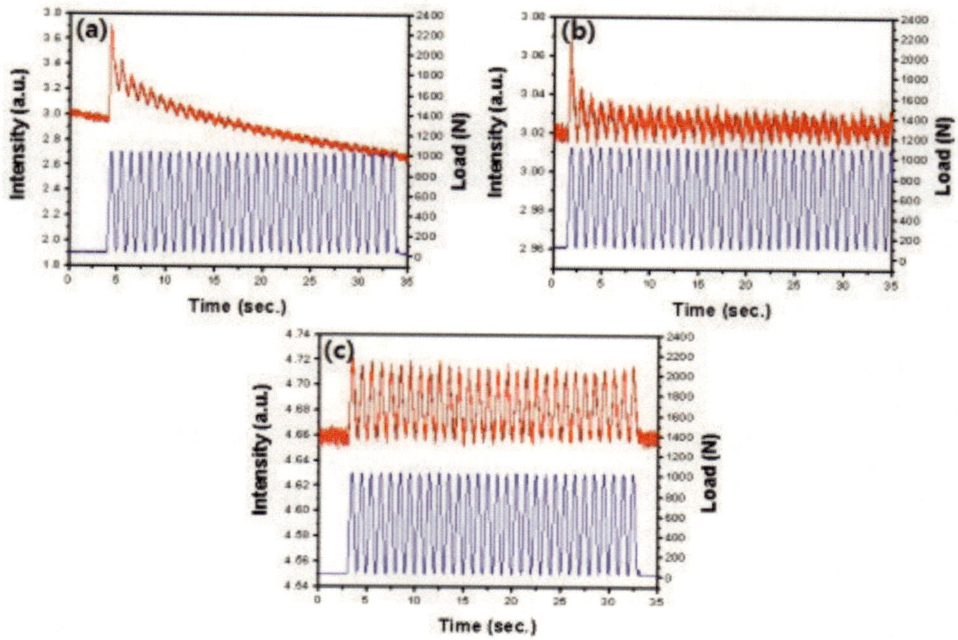

Figure 3.
TL intensity of SAOED response to the cyclic load at a frequency of 1 Hz under different irradiation conditions: (a) with UV irradiation turned off; (b) under a UV irradiation with a power density of 200 mW/cm2; (c) under a UV irradiation with a power density of 1000 mW/cm2. Reproduced by permission of the OSA Publishing [41].

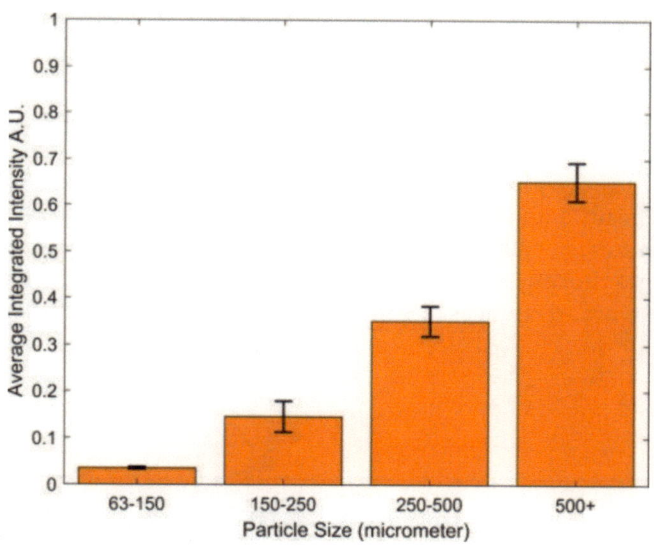

Figure 4.
TL integrated intensity on dependent of particle sizes of sucrose crystals. Reproduced by permission of the American Chemical Society [42].

3.4 External factor-dependent TL

The TL powders could be directly stimulated by ultrasonication or impact. The ultrasonic TL is dependent on the ultrasonic power with a linear relationship [43]. The impact-induced TL is strongly affected by the impact velocity or impact energy [2, 44]. When TL powders were composited in various matrices, other mechanical actions, such as rubbing, stretching, and compression, would be employed for TL. The intensity of the rubbing-induced TL shows relationships to both the applied normal load and the friction velocity [43]. For the stretching-induced TL, the elastic modulus plays a key role on the critical strain [45]. In addition, TL intensity varies along with the change of strain levels and stretching speeds [7]. For the compression-induced TL, it depends on the applied load as well as the deformation rate [31, 37].

4. Applications

TL materials could be composited in a variety of hosts, such as polymer matrices and metal bulk materials. The as-fabricated TL composites could emit light under the stimulus of mechanical behaviors for various applications. Because $SrAl_2O_4:Eu^{2+}$ and $ZnS:Mn^{2+}/Cu^{+}$ are the well-recognized intense TL materials, the present applications almost focus on them.

4.1 Structural health monitoring

The TL composites could be directly stimulated by the inner stress, showing application perspectives in structural health monitoring of devices, machines, and buildings [46–48]. To date, TL materials have been well employed to visualize and monitor the stress distribution as well as the fatigue crack initiation and propagation of matrices [1, 49, 50]. The sensitization of stress distribution in solids was first conducted by C-N. Xu et al. [1] They composited the green-emitting $SrAl_2O_4:Eu^{2+}$ TL powders in epoxy resins and confirmed that the TL behaviors of the $SrAl_2O_4:Eu^{2+}$/epoxy composites under a compressive load of 1000 N could reflect the stress distribution based on the experimental and simulative results. The $SrAl_2O_4:Eu^{2+}$/epoxy composites were further employed to realize the measurements of instantaneous R-curves and bridging stress in a fast-propagating crack system (**Figure 5**) [51].

Based on the above pioneering achievements, researchers successfully developed the structural health monitoring applications of TL materials in steel box girders [52], hydrogen storage cylinders [53], and gas pipelines [54]. Compared with the conventional monitoring methods by electrical and magnetic signals, the approach by TL signals shows advantages of contactless, wireless, convenient, and visualization [49, 52].

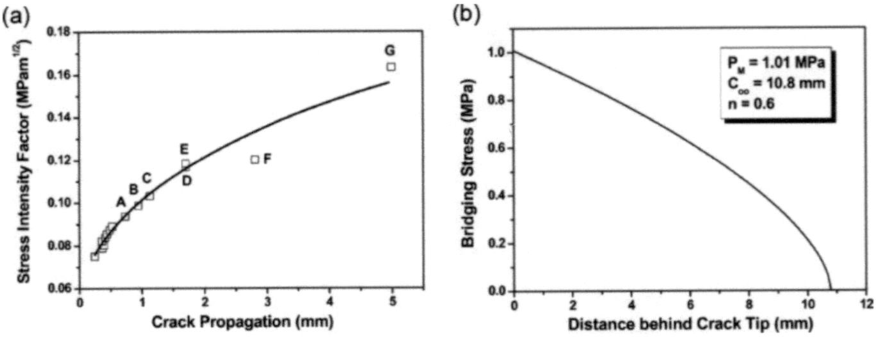

Figure 5.
Experimental R-curve and bridging stress distribution in the crack wake based on the TL of SAOED. Reproduced by permission of Elsevier B.V. [51].

4.2 Impact/load sensor

When TL materials undergo loading or impact, the emitted luminescence shows one-to-one correspondence between the emission intensity and impact/loading energy, which could be utilized to develop impact/load sensors to record the related mechanical information [2]. However, for the sensors fabricated from $SrAl_2O_4:Eu^{2+}$, the prominent problem is that the TL intensity will be decreased along with the increase of impact times or loading time, i.e., $SrAl_2O_4:Eu^{2+}$ shows poor cycling stability that goes against for its applications as impact/load sensors [41, 44]. Researchers further found that when an ultraviolet (UV) irradiation source with a certain power density was applied, $SrAl_2O_4:Eu^{2+}$ could keep the TL intensity stably based on the balance of trapping and de-trapping of the carriers in structure [41, 44]. The proposed $SrAl_2O_4:Eu^{2+}$-based sensor under UV irradiation could stably sensitize the applied load both in dynamic and static states (**Figure 6**) [41]. Differing from $SrAl_2O_4:Eu^{2+}$, ZnS:Mn/Cu showed almost no TL intensity degradation along with the increase of cycle numbers because of the piezoelectric effect, which could be directly used for impact/load sensor applications without needing an extra UV irradiation [37, 55].

4.3 Lighting, imaging, and displaying

The exploited devices for lighting, imaging, and displaying are mainly fabricated from ZnS:Mn/Cu and elastomer matrices. The as-fabricated ZnS:Cu/PDMS flexible composites showed bright and durable TL under stretching with a brightness of ca. 120 cd/m^2 and durability over 10,000 cycles [38]. The composites could be further fabricated into fabrics with patterns that could be applied for imaging and displaying as presented in **Figure 7** [56].

In addition to the stimulus of stretching and rubbing, the TL composites could also be activated by various mechanical sources, such as wind [5], magnetic field [57], and ultrasonic wave [58], which fulfill the requirements of green and sustainable developments. For practical applications in lighting and displaying, the TL flexible devices with a white light or multicolored emissions are required, and a variety of strategies have been proposed. For example, Jeong et al. employed ZnS:Cu, Mn and ZnS:Cu as the orange and green TL materials, respectively, and fabricated ZnS-based flexible composites, in which TL color manipulation including a warm white light was demonstrated by adjusting the component ratios of

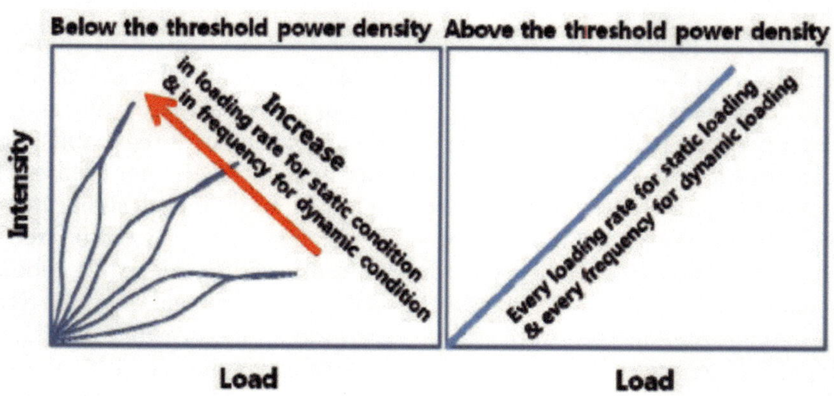

Figure 6.
Load responsiveness of SAOED in static and dynamic states with UV irradiation turned on and turned off. Reproduced by permission of the OSA Publishing [41].

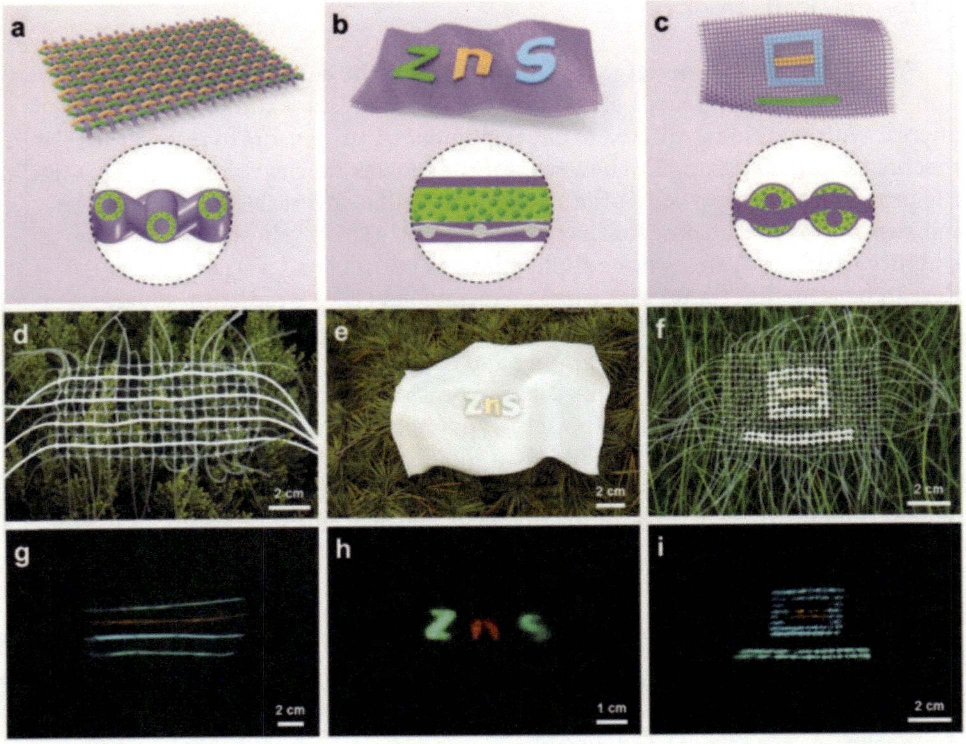

Figure 7.
TL fabrics based on the doped ZnS (a) fibers, (b) ribbons, and (c) dots; corresponding (d–f) optical and (g–i) TL photographs of the fabrics in (a–c). Reproduced by permission of the Royal Society of Chemistry [56].

ZnS:Cu, Mn and ZnS:Cu [36]. They further presented a strategy for the TL color manipulation of doped ZnS by physically combining fluorescent dyes in PDMS elastomers based on the energy transfer between the TL of doped ZnS and the PL of dyes [59]. Hao and his co-workers also realized the remote tuning of TL color of ZnS:Al, Cu/PDMS composites by modulating the frequency of magnetic field [60]. In addition, flexible devices with dual-mode emissions, i.e., EL and TL, have also been developed for imaging and displaying [59, 61].

4.4 Pressure sensor

The TL flexible composites exhibit luminescent signals dependent on the applied pressure. Based on such performance, Wang et al. developed a ZnS:Mn-based pressure sensor for both single-point dynamic pressure recording and 2D planar pressure mapping with a high spatial resolution of 100 μm and a fast response time less than 10 ms [24]. The pressure sensor was further used as a flexible handwriting device that could collect the information of both signatures and signing habits as shown in **Figure 8**, exhibiting high-level security compared with the existing technologies. They further introduced the single-electrode triboelectric nanogenerator in the ZnS:Mn-based flexible composites and obtained a full dynamic-range pressure sensor for the visualization of pressure distribution both in low pressure regimes (< 100 kPa) and high-pressure regimes (> 1 MPa) with an excellent pressure sensitivity of 6 MPa^{-1} [62]. In addition, CaZnOS:Er^{3+} thin-film was prepared, which possessed the pressure and temperature sensing based on its TL and upconversion luminescence [63].

Figure 8.
Flexible handwriting device based on the TL of ZnS:Mn for visualization of dynamic pressure distributions: (a) schematic illustration of the system; (b) visualization of 2D planar pressure distribution; (c-h) visualization of the signing process. Reproduced by permission of Wiley-VCH [24].

4.5 Stress/strain sensor

When the TL materials are introduced in elastic matrices, stress/strain sensor could be obtained. At present, the widely employed TL materials for fabricating stress/strain sensors are ZnS:Mn, ZnS:Cu, and $SrAl_2O_4$:Eu, because of their prominent TL properties as well as the one-to-one correspondence between the TL intensity and stress/strain. Yun et al. [64] further found that the co-dopant of Dy^{3+} in $SrAl_2O_4$:Eu could improve its performance as stress sensor based on the sensitivity. In addition to sense the stress or strain by analyzing the TL intensity, the risetime and decay time of TL during cyclic elastic deformation of $SrAl_2O_4$:Eu were also demonstrated to be suitable for evaluating the change of the strain energy [65]. Moreover, a calibration method for $SrAl_2O_4$:Eu, Dy-based thin-film sensor was proposed to enable quantitative full-field strain measurements in pixel-level resolution [66]. Qian et al. [45] prepared ZnS:Mn/Cu@Al_2O_3/PDMS flexible composites and adjusted the elastic modulus by introducing SiO_2 nanoparticles. They finally obtained a TL stress/strain sensor that could be driven by weak mechanics of skin movements.

In the very recent work [7], $Sr_3Al_2O_6$:Eu with bright and tunable PL and TL was presented when it was composited in PDMS elastomers. By combining the wavelength selectivity of PL and dynamic stress responsiveness of TL, a multimode stretching/strain sensor was developed by a bilayered structure design of $Sr_3Al_2O_6$:Eu/PDMS composites with coating a light-shielding layer of Au atop (as shown in **Figure 9**). The fabricated sensor could sense the stretching states and strain levels simultaneously, breaking the limit of static strain sensing in previous researches.

4.6 Mechanics-light-electricity conversion

The TL materials could convert mechanics into light, which could be further utilized to generate electricity for various applications. When the $SrAl_2O_4$:Eu/epoxy TL composites were combined in a commercial silicon solar cell, the mechanics-light-electricity conversion could be achieved [67]. In addition to the generation of electricity by utilizing the mechanics-induced luminescence, TL materials could be combined with a nanogenerator and convert the input mechanical stimuli to electric and light simultaneously [68]. The TL materials could also be composited with a photocatalyst to realize the catalysis activity in dark under the stimuli of mechanics [69]. The above conversion systems based on TL show great perspectives for applications in dark environments, such as deep sea and polar night region.

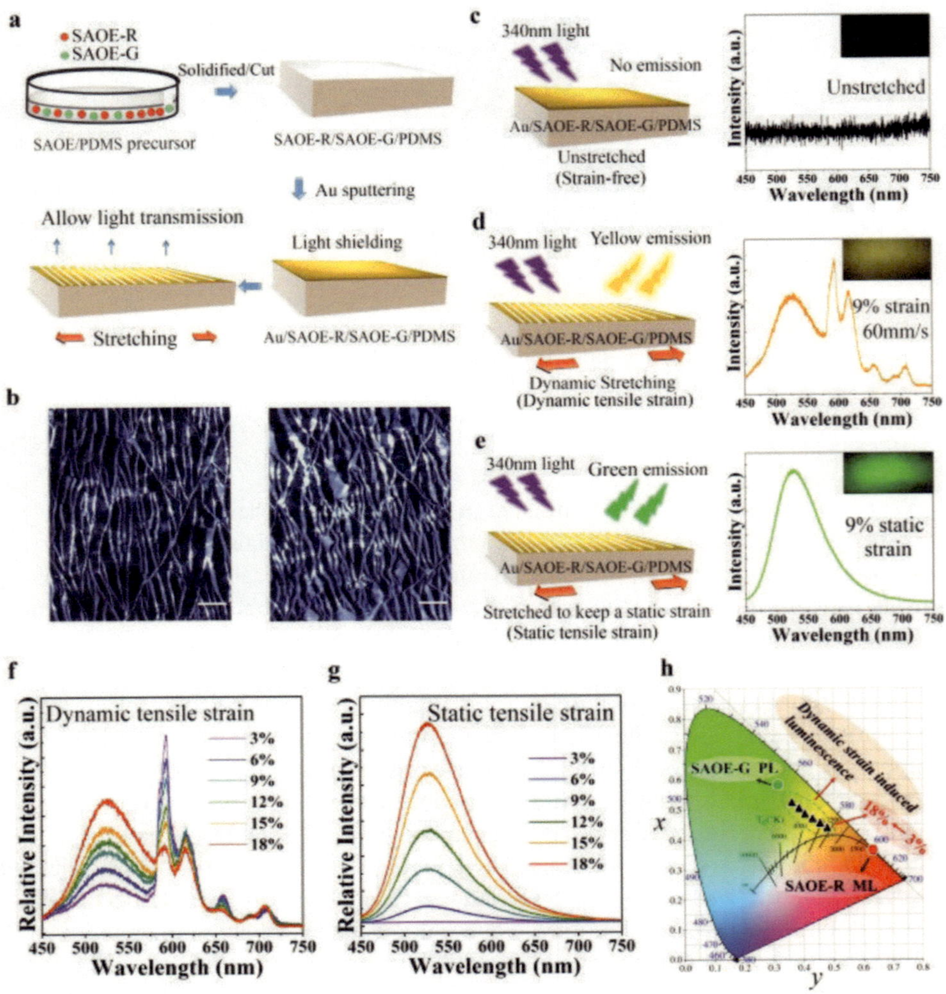

Figure 9.
Multimode stretching/strain sensor based on the TL and PL of Sr3Al2O6:Eu: (a) fabricating process; (b) crack opening when stretched, scale bar: 100 μm; (c–e) stretching state responses; (f–g) strain level responses; (h) corresponding color conversion based on various dynamic strain levels. Reproduced by permission of Wiley-VCH [7].

4.7 Biological applications

Because some of the TL materials show good biocompatibility, such as the rare earth-doped oxide ceramics, they are promising for the detection of mechanical behaviors in biological tissues/organs. The $SrAl_2O_4$:Eu TL powders was applied in the synthetic bone, and the related mechanical dynamic environment was monitored with a high-definition and high-speed visualization [70]. SAOED powders were also applied in artificial tooth for occlusal examination [71]. The composition of SAOED in the commercial denture base resin (DBR) could not only endow with bright TL but also improve its mechanical performance. As a result, an artificial tooth model with SAOED was made in which bright and sensitive TL could be directly observed to guide clinicians to purposefully adjust the occlusal surface until a balanced occlusion established.

5. Conclusions

In summary, we present a comprehensive overview on the study of TL. The material systems in both organics and inorganics, unique spectral characteristics, and TL performance, as well as the representative applications in various fields, are included. We hope that this chapter could help researchers in the field to gain a comprehensive and in-depth understanding of TL and stimulate continued interests and endeavors in this area to promote more innovative applications.

Acknowledgements

The authors thank the support from the Natural Science Foundation of Gansu Province (17JR5RA319) and the CAS Pioneer Hundred Talents Program.

Conflict of interest

The authors declare no conflict of interest.

Author details

Zhaofeng Wang* and Fu Wang
State Key Laboratory of Solid Lubrication, Lanzhou Institute of Chemical Physics, Chinese Academy of Sciences, Lanzhou, Gansu, China

*Address all correspondence to: zhfwang@licp.cas.cn

References

[1] Xu C-N, Watanabe T, Akiyama M, Zheng X-G. Direct view of stress distribution in solid by mechanoluminescence. Applied Physics Letters. 1999;**74**:2414-2416

[2] Jha P, Khare A, Singh PK, Chandra VK, Sonwane VD. Ball impact induced elastico-mechanoluminescence for impact sensor. Journal of Luminescence. 2018;**195**:40-43

[3] Wang X, Xu C-N, Yamada H, Nishikubo K, Zheng X-G. Electro-mechano-optical conversions in Pr^{3+}-doped $BaTiO_3$–$CaTiO_3$ ceramics. Advanced Materials. 2005;**17**:1254-1258

[4] Feng A, Smet PF. A review of mechanoluminescence in inorganic solids: Compounds, mechanisms, models and applications. Materials. 2018;**11**:484

[5] Jeong SM, Song S, Joo K-I, Kim J, Hwang S-H, Jeong J, et al. Bright, wind-driven white mechanoluminescence from zinc sulphide microparticles embedded in a polydimethylsiloxane elastomer. Energy & Environmental Science. 2014;**7**:3338-3346

[6] Dengfeng P, Bing C, Feng W. Recent advances in doped mechanoluminescent phosphors. ChemPlusChem. 2015;**80**:1209-1215

[7] Wu C, Zeng S, Wang Z, Wang F, Zhou H, Zhang J, et al. Efficient mechanoluminescent elastomers for dual-responsive anticounterfeiting device and stretching/strain sensor with multimode sensibility. Advanced Functional Materials. 2018:1803168

[8] Olawale DO, Dickens T, Sullivan WG, Okoli OI, Sobanjo JO, Wang B. Progress in triboluminescence-based smart optical sensor system. Journal of Luminescence. 2011;**131**:1407-1418

[9] Zhang H, Yamada H, Terasaki N, Xu C-N. Blue light emission from stress-activated $CaYAl_3O_7$:Eu. Journal of the Electrochemical Society. 2008;**155**:J128-J131

[10] Zhang J-C, Zhao L-Z, Long Y-Z, Zhang H-D, Sun B, Han W-P, et al. Color Manipulation of intense multiluminescence from CaZnOS:Mn^{2+} by Mn^{2+} concentration effect. Chemistry of Materials. 2015;**27**:7481-7489

[11] Akiyama M, Xu C-N, Liu Y, Nonaka K, Watanabe T. Influence of Eu, Dy co-doped strontium aluminate composition on mechanoluminescence intensity. Journal of Luminescence. 2002;**97**:13-18

[12] Xie Y, Li Z. Triboluminescence: Recalling interest and new aspects. Chem. 2018;**4**:943-971

[13] Zink JI, Hardy GE, Sutton JE. Triboluminescence of sugars. The Journal of Physical Chemistry. 1976;**80**:248-249

[14] Chandra BP. Mechanoluminescence and piezoelectric behaviour of molecular crystals. Physica Status Solidi. 1981;**64**:395-405

[15] Chandra BP, Zink JI. Triboluminescence of triclinic crystals. Journal of Luminescence. 1981;**23**:363-372

[16] Hardy GE, Baldwin JC, Zink JI, Kaska WC, Liu P-H, Dubois L. Triboluminescence spectroscopy of aromatic compounds. Journal of the American Chemical Society. 1977;**99**:3552-3558

[17] Sweeting LM, Rheingold AL. Crystal structure and triboluminescence. 1. 9-Anthryl carbinols. The Journal of Physical Chemistry. 1988;**92**:5648-5655

[18] Wang C, Xu B, Li M, Chi Z, Xie Y, Li Q, et al. A stable tetraphenylethene derivative: Aggregation-induced emission, different crystalline polymorphs, and totally different mechanoluminescence properties. Materials Horizons. 2016;**3**:220-225

[19] Fontenot RS, Hollerman WA, Bhat KN, Aggarwal MD. Synthesis and characterization of highly triboluminescent doped europium tetrakis compounds. Journal of Luminescence. 2012;**132**:1812-1818

[20] Marchetti F, Di Nicola C, Pettinari R, Timokhin I, Pettinari C. Synthesis of a photoluminescent and triboluminescent copper(I) compound: An experiment for an advanced inorganic chemistry laboratory. Journal of Chemical Education. 2012;**89**:652-655

[21] Butler CT. Room-temperature deformation luminescence in alkali halides. Physical Review. 1966;**141**:750-757

[22] Rai RK, Upadhyay AK, Kher RS, Dhoble SJ. Mechanoluminescence, thermoluminescence and photoluminescence studies on Al_2O_3:Tb phosphors. Journal of Luminescence. 2012;**132**:210-214

[23] Akiyama M, Xu C-N, Nonaka K. Intense visible light emission from stress-activated ZrO_2:Ti. Applied Physics Letters. 2002;**81**:457-459

[24] Wang X, Zhang H, Yu R, Dong L, Peng D, Zhang A, et al. Dynamic pressure mapping of personalized handwriting by a flexible sensor matrix based on the mechanoluminescence process. Advanced Materials. 2015;**27**:2324-2331

[25] Li L, Wong K-L, Li P, Peng M. Mechanoluminescence properties of Mn^{2+}-doped BaZnOS phosphor. Journal of Materials Chemistry C. 2016;**4**:8166-8170

[26] Zhang H, Terasaki N, Yamada H, Xu C-N. Mechanoluminescence of europium-doped $SrAMgSi_2O_7$ (A=Ca, Sr, Ba). Japanese Journal of Applied Physics. 2009;**48**:04C109

[27] Sahu AK, Kore BP, Chowdhary PS, Nayar V, Dhoble SJ. Systematic study of photoluminescence, lyoluminescence and mechanoluminescence in Ce^{3+}- and Eu^{3+}-activated Li_3PO_4 phosphors. Luminescence. 2014;**29**:58-64

[28] Kamimura S, Yamada H, Xu C-N. Development of new elasticoluminescent material $SrMg_2(PO_4)_2$:Eu. Journal of Luminescence. 2012;**132**:526-530

[29] Mishra GC, Upadhyay AK, Dwiwedi SK, Dhoble SJ, Kher RS. Correlation between thermoluminescence and mechanoluminescence of γ-ray-irradiated Dy-doped BaB_4O_7 phosphors. Journal of Materials Science. 2012;**47**:2752-2756

[30] Zhang J-C, Long Y-Z, Wang X, Xu C-N. Controlling elastico-mechanoluminescence in diphase $(Ba,Ca)TiO_3$:Pr^{3+} by co-doping different rare earth ions. RSC Advances. 2014;**4**:40665-40675

[31] Zhang J-C, Long Y-Z, Yan X, Wang X, Wang F. Creating recoverable mechanoluminescence in piezoelectric calcium niobates through Pr^{3+} doping. Chemistry of Materials. 2016;**28**:4052-4057

[32] Dong T, Chao-Nan X, Akihito Y, Masayoshi F, Jou H, Xu-Guang Z. LiNbO3:Pr3+: A multipiezo material with simultaneous piezoelectricity and sensitive piezoluminescence. Advanced Materials. 2017;**29**:1606914

[33] Kamimura S, Yamada H, Xu C-N. Purple photochromism in Sr_2SnO_4:Eu^{3+} with layered perovskite-related structure. Applied Physics Letters. 2013;**102**:031110

[34] Upadhyay AK, Dhoble SJ, Kher RS. Mechanoluminescence properties of gamma-ray-irradiated $BaSO_4$:Eu phosphors. Luminescence. 2011;**26**:471-476

[35] Botterman J, KVd E, Baere ID, Poelman D, Smet PF. Mechanoluminescence in $BaSi_2O_2N_2$:Eu. Acta Materialia. 2012;**60**:5494-5500

[36] Moon JS, Seongkyu S, Soo-Keun L, Young HN. Color manipulation of mechanoluminescence from stress-activated composite films. Advanced Materials. 2013;**25**:6194-6200

[37] Shohag MAS, Tran SA, Ndebele T, Adhikari N, Okoli OI. Designing and implementation of triboluminescent materials for real-time load monitoring. Materials and Design. 2018;**153**:86-93

[38] Jeong SM, Song S, Lee S-K, Choi B. Mechanically driven light-generator with high durability. Applied Physics Letters. 2013;**102**:051110

[39] Sohn K-S, Park WB, Timilsina S, Kim JS. Mechanoluminescence of $SrAl_2O_4$:Eu^{2+}, Dy^{3+} under cyclic loading. Optics Letters. 2014;**39**:1410-1413

[40] Fan X-H, Zhang J-C, Zhang M, Pan C, Yan X, Han W-P, et al. Piezoluminescence from ferroelectric $Ca_3Ti_2O_7$:Pr^{3+} long-persistent phosphor. Optics Express. 2017;**25**:14238-14246

[41] Sohn K-S, Cho MY, Kim M, Kim JS. A smart load-sensing system using standardized mechano-luminescence measurement. Optics Express. 2015;**23**:6073-6082

[42] Smith CJ, Griffin SR, Eakins GS, Deng F, White JK, Thirunahari S, et al. Triboluminescence from pharmaceutical formulations. Analytical Chemistry. 2018;**90**:6893-6898

[43] Zhang J-C, Xu C-N, Kamimura S, Terasawa Y, Yamada H, Wang X. An intense elastico-mechanoluminescence material CaZnOS:Mn^{2+} for sensing and imaging multiple mechanical stresses. Optics Express. 2013;**21**:12976-12986

[44] Mao Q, Chen Z, Ji Z, Xi J. UV-assisted mechanoluminescence properties of $SrAl_2O_4$: (Eu,Dy) for impact sensing. Journal of Materials Science. 2017;**52**:8370-8376

[45] Xin Q, Zheren C, Meng S, Fengyu L, Wei F, Yudong L, et al. Printable skin-driven mechanoluminescence devices via nanodoped matrix modification. Advanced Materials. 2018;**30**:1800291

[46] Krishnan S, Van der Walt H, Venkatesh V, Sundaresan VB. Dynamic characterization of elastico-mechanoluminescence towards structural health monitoring. Journal of Intelligent Material Systems and Structures. 2017;**28**:2458-2464

[47] Suman T, Ho LK, Nam KY, Sik KJ. Optical evaluation of in situ crack propagation by using mechanoluminescence of $SrAl_2O_4$:Eu^{2+}, Dy^{3+}. Journal of the American Ceramic Society. 2015;**98**:2197-2204

[48] Scheiner M, Hammel E, Okoli OI. Ultraviolet priming of triboluminescence. Journal of Luminescence. 2018;**194**:803-805

[49] Xu C-N, Zheng X-G, Akiyama M, Nonaka K, Watanabe T. Dynamic visualization of stress distribution by mechanoluminescence image. Applied Physics Letters. 2000;**76**:179-181

[50] Terasaki N, Xu CN. Historical-log recording system for crack opening and growth based on mechanoluminescent flexible sensor. IEEE Sensors Journal. 2013;**13**:3999-4004

[51] Kim JS, Kwon Y-N, Sohn K-S. Dynamic visualization of crack

propagation and bridging stress using the mechano-luminescence of $SrAl_2O_4$: (Eu,Dy,Nd). Acta Materialia. 2003;**51**:6437-6442

[52] Yoshida A, Liu L, Tu D, Kainuma S, Xu C-N. Mechanoluminescent testing as an efficient inspection technique for the management of infrastructures. Journal of Disaster Research. 2017;**12**:506-514

[53] Fujio Y, Xu C-N, Terasawa Y, Sakata Y, Yamabe J, Ueno N, et al. Sheet sensor using $SrAl_2O_4$:Eu mechanoluminescent material for visualizing inner crack of high-pressure hydrogen vessel. International Journal of Hydrogen Energy. 2016;**41**:1333-1340

[54] Yang Y, Zheng S, Fu X, Zhang H. Remote and portable mechanoluminescence sensing system based on a $SrAl_2O_4$:Eu,Dy film and its potential application to monitoring the safety of gas pipelines. Optik. 2018;**158**:602-609

[55] Shohag MAS, Jiang Z, Hammel EC, Braga Carani L, Olawale DO, Dickens TJ, et al. Development of friction-induced triboluminescent sensor for load monitoring. Journal of Intelligent Material Systems and Structures. 2018;**29**:883-895

[56] Zhang J, Bao L, Lou H, Deng J, Chen A, Hu Y, et al. Flexible and stretchable mechanoluminescent fiber and fabric. Journal of Materials Chemistry C. 2017;**5**:8027-8032

[57] Wong M-C, Chen L, Tsang M-K, Zhang Y, Hao J. Magnetic-induced luminescence from flexible composite laminates by coupling magnetic field to piezophotonic effect. Advanced Materials. 2015;**27**:4488-4495

[58] Terasaki N, Yamada H, Xu C-N. Ultrasonic wave induced mechanoluminescence and its application for photocatalysis as ubiquitous light source. Catalysis Today. 2013;**201**:203-208

[59] Moon JS, Seongkyu S, Hyunmin K, Kyung-Il J, Hideo T. Mechanoluminescence color conversion by spontaneous fluorescent-dye-diffusion in elastomeric zinc sulfide composite. Advanced Functional Materials. 2016;**26**:4848-4858

[60] Wong M-C, Chen L, Bai G, Huang L-B, Hao J. Temporal and remote tuning of piezophotonic-effect-induced luminescence and color gamut via modulating magnetic field. Advanced Materials. 2017;**29**:1701945

[61] Jeong SM, Song S, Kim H. Simultaneous dual-channel blue/green emission from electro-mechanically powered elastomeric zinc sulphide composite. Nano Energy. 2016;**21**:154-161

[62] Wang X, Que M, Chen M, Han X, Li X, Pan C, et al. Full dynamic-range pressure sensor matrix based on optical and electrical dual-mode sensing. Advanced Materials. 2017;**29**:1605817

[63] Zhang H, Peng D, Wang W, Dong L, Pan C. Mechanically induced light emission and infrared-laser-induced upconversion in the Er-doped CaZnOS multifunctional piezoelectric semiconductor for optical pressure and temperature sensing. The Journal of Physical Chemistry C. 2015;**119**:28136-28142

[64] Gun Jin Y, Mohammad Reza R, Amir Hossein G, Gong-Cheol L, Jun-Seong C. Stress sensing performance using mechanoluminescence of $SrAl_2O_4$:Eu (SAOE) and $SrAl_2O_4$:Eu, Dy (SAOED) under mechanical loadings. Smart Materials and Structures. 2013;**22**:055006

[65] Someya S, Ishii K, Munakata T, Saeki M. Lifetime-based measurement of stress during

cyclic elastic deformation using mechanoluminescence of $SrAl_2O_4:Eu^{2+}$. Optics Express. 2014;**22**:21991-21998

[66] Ali Imani A, Mohammad Reza R, Gun Jin Y. Quantitative full-field strain measurements by SAOED ($SrAl_2O_4:Eu^{2+}$, Dy^{3+}) mechanoluminescent materials. Smart Materials and Structures. 2016;**25**:095032

[67] Terasaki N, Xu C-N, Imai Y, Yamada H. Photocell system driven by mechanoluminescence. Japanese Journal of Applied Physics. 2007;**46**:2385

[68] Huajing F, Xiandi W, Qiang L, Dengfeng P, Qingfeng Y, Caofeng P. A stretchable nanogenerator with electric/light dual-mode energy conversion. Advanced Energy Materials. 2016;**6**:1600829

[69] Terasaki N, Zhang H, Imai Y, Yamada H, Xu C-N. Hybrid material consisting of mechanoluminescent material and TiO_2 photocatalyst. Thin Solid Films. 2009;**518**:473-476

[70] Hyodo K, Xu C, Mishima H, Miyakawa S. Optical Stress Imaging for Orthopedic Biomechanics-Comparison of Thermoelastic Stress Analysis and Developed Mechanoluminescent Method. Berlin, Heidelberg: Springer Berlin Heidelberg; 2010. pp. 545-548

[71] Jiang Y, Wang F, Zhou H, Fan Z, Wu C, Zhang J, et al. Optimization of strontium aluminate-based mechanoluminescence materials for occlusal examination of artificial tooth. Materials Science and Engineering: C. 2018;**92**:374-380

Chapter 3

Mechanism of Photoluminescence in Erbium-Doped Chalcogenide

Volodymyr V. Halyan and Inna A. Ivashchenko

Abstract

The monograph describes the technique of the synthesis of glasses and the method of the growth of erbium-doped single crystals. The photoluminescence spectra of $Ag_{0.05}Ga_{0.05}Ge_{0.95}S_2$-$Er_2S_3$ glasses and glasses from the Ga_2S_3-La_2S_3-Er_2S_3 system have been investigated in the visible and near-infrared ranges. According to the energy transitions in the erbium ions, a radiation mechanism for conversion and up-conversion luminescence has been established. The role of structural ordering and the influence of defects on the radiation efficiency of Er^{3+} ions have been investigated. The spectra of photoluminescence of $(Ga_{54.59}In_{44.66}Er_{0.75})_2S_{300}$ and $(Ga_{69.75}La_{29.75}Er_{0.5})_2S_{300}$ single crystals have been studied. The efficiency of the radiation of the amorphous and crystalline materials has been compared. Also, the temperature dependence of the integral intensity of the radiation of glasses and single crystals has been studied. It is established that in a limited temperature range, these materials can be used for the manufacture of non-contact optical thermosensors.

Keywords: glass, single crystal, photoluminescence, energy transition, radiation mechanism, visible, and near-infrared ranges

1. Introduction

Modern scientific and technological progress requires the constant search for and introduction of new multifunctional materials. One of the directions in the search for new semiconductor materials is the study of multicomponent systems of binary and ternary phases, which are promising for practical use. This is due to increasing requirements for semiconductor materials in connection with the development of new or modernization of known technical devices. To expand the list of materials, a study of chalcogenide quasi-ternary systems was undertaken.

Chalcogenide semiconductor materials were the object of research for many decades. The best studied glassy and crystalline chalcogenides are based on the Ge-X, Ga-X, As-X systems (X = S, Se, Te) modified by the admixture of Hg, Sb, Pb, Ag, etc. [1–3]. An important feature of chalcogenide glass is that the introduced admixtures are electrically and optically inactive. As a rule, they do not create localized states in the band gap. When introduced into a glass-forming matrix, admixture atoms rebuild their local environment resulting in the saturation of their valence bonds. Exceptions are the admixtures of bismuth, platinum, and gold into arsenic sulfide [4–6].

Special attention is devoted to the research of the luminescence properties of chalcogenide crystals and glasses due to practical application in optoelectronic technology. Their unique properties create advantages over other light-emitting materials and cause considerable interest both from the fundamental and the applied point of view. The chalcogenide materials combine high transparency in the visible, near-infrared, and medium-infrared spectral regions. By selecting the optimal component composition, it is possible to obtain wide areas of glass formation and to introduce relatively high concentration of rare earth metals (RE). Additionally, chalcogenides are characterized by high refractive index, good nonlinear optical properties, resistance to aggressive media, and easy manufacture technology. The addition of RE (Er, La, Eu, Pr, Tb, Ho, etc.) to the crystalline or amorphous chalcogenide medium creates potential opportunities for their use as high-quantum output luminophores, optical filters, active media in laser technology, in telecommunication devices, as well as non-contact optical temperature, and γ-irradiation sensors. The chalcogenide media that are activated by RE ions can also exhibit both conversion and up-conversion photoluminescence (PL) which creates the prerequisites for the design of effective converters from the infrared range to visible light.

Therefore, the preparation of chalcogenide glasses and crystals doped with rare earths and the study of their emission properties under different temperature and irradiation regimes are an extremely important area of modern solid-state physics and chemistry.

2. Synthesis of glasses and single crystal growth

Elementary high-purity substances were used for the synthesis of samples: Ag (99.99 wt.% of the principal), Ga, In, Ge (99.999 wt.% purity), La, Er (99.9 wt.%), and S, Se (99.997 wt.%). The elements for single crystal growth were further purified by double vacuum distillation. The weight of the starting components for the glasses was 3 g; the batches for the synthesis of single crystals were 10 g.

The samples were synthesized in cylindrical quartz containers with 9–15 mm diameter. The loaded ampoules were evacuated to a residual pressure of 1.33×10^{-2} Pa. The synthesis of glasses was performed in a shaft-type furnace with a temperature control system of ± 5 K accuracy.

The alloys for the glasses of the La_2S_3-Er_2S_3-Ga_2S_3 system with sample weight 2 g were pre-synthesized at 870 K and 24 h exposure. Obtained samples were ground in an agate mortar and loaded into quartz containers with a spherical bottom of 1 cm diameter. The heating was stepwise, first to 1070 K at a rate of 50 K/h and then to 1420 K at a rate of 40 K/h. After 3 h exposure at the maximum temperature, the samples were quenched into saturated NaCl solution with crushed ice.

Synthesis of the $Ag_{0.05}Ga_{0.05}Ge_{0.95}S_2$-$Er_2S_3$ glass alloys was performed in two stages. Initially, the ampoules were heated in the flame of an oxygen-gas burner to complete bonding of elemental sulfur. Then the ampoules were placed in a shaft-type furnace and heated at a rate of 20 K/h to 1273 K. The samples were kept at the maximum temperature for 10 h with periodic vibration. The alloys were cooled at a rate of 10–20 K/h to annealing temperature. The homogenizing annealing was held for 500 h at 720 K. The annealed alloys were quenched into 25% aqueous NaCl solution at room temperature.

To prevent the spatter of melt during quenching as well as to reduce losses due to the vapor phase condensing on the walls of the container, the special form of the container was used (**Figure 1**), and the upper part was thermostated by asbestos cord after the binding of sulfur in the oxygen-gas burner flame was used.

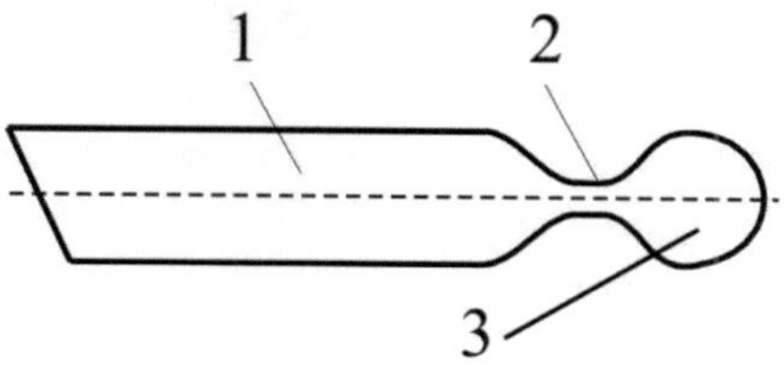

Figure 1.
Container for the synthesis of glassy alloys: (1) cylindrical part of the container; (2) neck; (3) thin-walled pear-shaped chamber.

The method and conditions for the single crystal growth of the phases $(Ga_{55}In_{45})_2S_{300}$ and $(Ga_{55}In_{45})_2S_{300}$:0.3 at.% Er [7] were selected from the Ga_2S_3-In_2S_3 phase diagram; the supercooling temperature was determined from the cooling curves of the sample thermograms. The solution-melt method was used; supercooling of the solution melt was 70 K. The synthesis of the initial alloys at the maximum temperature of 1200 K and the growth of crystals were combined in evacuated graphitized quartz container with a conical bottom and a 2 mm diameter neck. The growth process was performed in a vertical two-zone furnace. The maximum temperature was 1200 K; the temperature gradient at the solid-melt interface was 20 K/cm. After melting the batch, the ampoule was lowered at the maximum rate. After crystallization of 10 mm of melt along the length of the ampoule, the growth was stopped, followed by re-melting of 6.0–8.0 mm of the crystallized portion and by annealing the seed during 100 h. Further growth of the single crystal was performed at a rate of 5 mm/day. After completion of the process, both furnaces were cooled to 820 K at a rate of 50–70 K/day, and the resulting single crystal

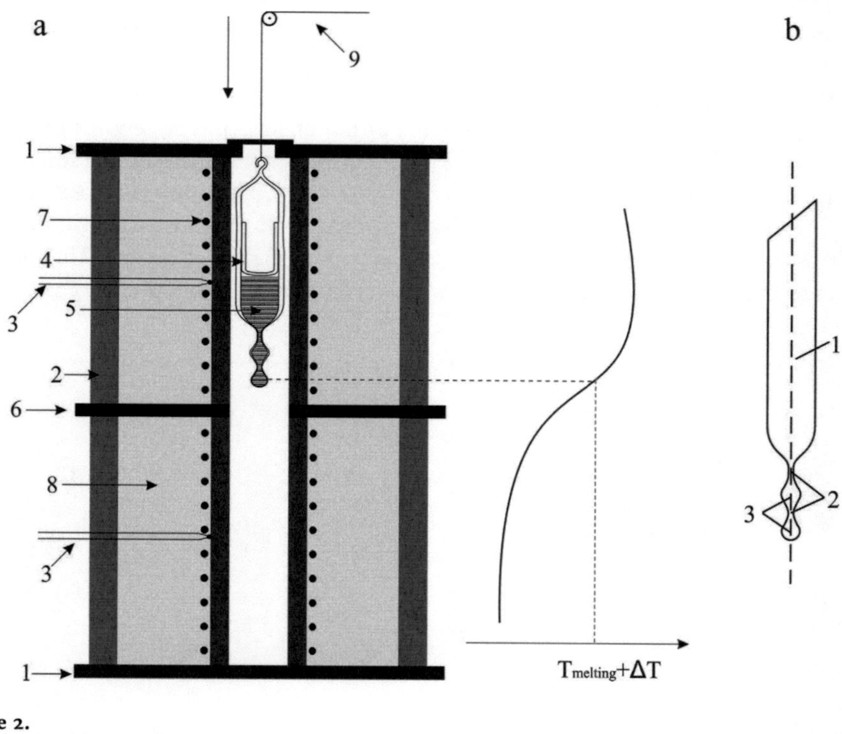

Figure 2.
Growth unit setup and the container for the single crystal growth: (a) 1—metal flange, 2—asbestos-cement casing, 3—Pt/Pt-Rh-thermocouple, 4—growth container, 5—melt, 6—metal disk, 7—heater, 8—thermal insulator, and 9—pulley for moving ampoules; (b) 1—cylindrical part of the container, 2—necks for seed formation, and 3—pear-shaped chambers.

was annealed for 100 h. After that, the furnace was switched off. The single crystals of orange color, 14 mm in diameter and 20 mm in length, were obtained.

The single crystal growth was performed in a setup installation shown in **Figure 2a** along with the temperature distribution along the heaters. The temperature in the growth zone was maintained by precision temperature controllers with an accuracy of ±0.5 K.

A version of a container with two pear-shaped chambers connected by a neck was used for the growth of single crystals (**Figure 2b**). This increases the probability of obtaining a single crystal with its subsequent growth to a larger size in the cylindrical portion of the container.

Independent temperature control in different areas of the heater allows us to vary the gradient at the solid-melt interface within 3–5 K/mm. The growth rate was within 2 mm/day. The vertical movement of the plane of the crystallization zone was ensured by moving the container while the heater was at a fixed position.

3. Photoluminescence in chalcogenide glasses

An analysis of literature sources shows that chalcogenide glasses, unless doped with REs, do not usually exhibit luminescent properties at room temperature. For a long time, this limited their use in optoelectronic technology. Investigations of RE-free glasses with the substitution of selenium for sulfur revealed the existence of a broad unstructured PL band at 80 K (**Figure 3**).

The glasses of the $AgGaSe_2 + GeS_2 \Leftrightarrow AgGaS_2 + GeSe_2$ system excited by a laser with 532-nm wavelength exhibit at low-temperature luminescence with a single maximum in the near-infrared range ($\lambda_m \approx 1150$–1180 nm) with an emission band half-width $\Delta E \approx 0.26$–0.30 eV (**Figure 3a**), which is typical of the recombination luminescence in disordered systems. The intensity of the luminescence maximum depends on the content of Se (when sulfur is substituted with selenium, **Figure 3b**). The dependence is complex and to a large extent is due to the change in the level of luminescence excitation (caused by the shift of the absorption edge with increasing concentration of selenium in the alloy [8]). An important feature of the luminescence spectrum is that the position of the luminescence maximum is close to the center of the band gap (as determined by the position of optical absorption edge [8]). This is well illustrated by the model of Mott and Davis [9] stating that a band of localized states (several tenths of eV wide) is located in the center of the band

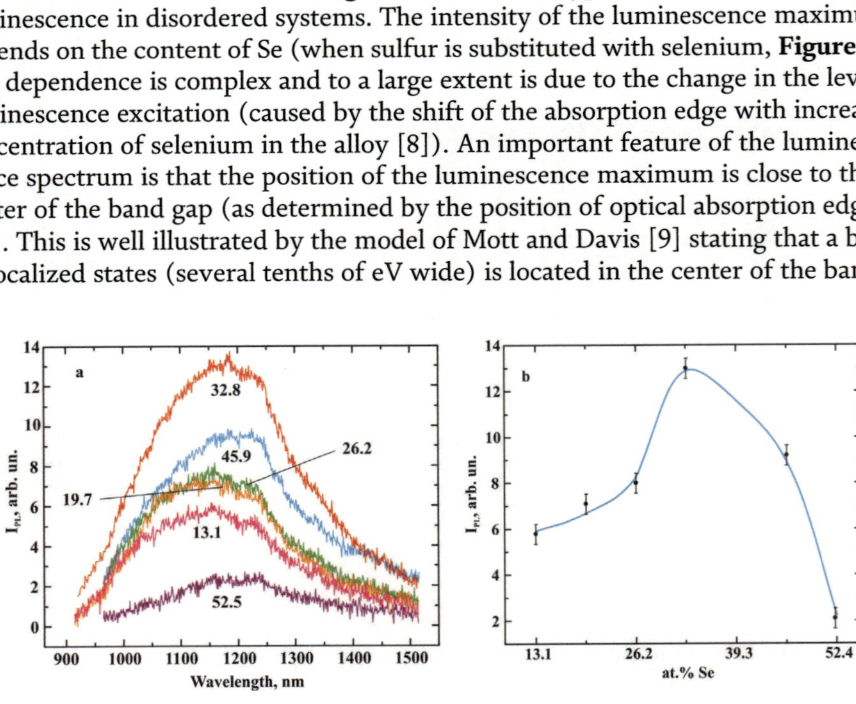

Figure 3.
(a) Luminescence spectra of the glassy alloys of the $AgGaSe_2 + GeS_2$ $AgGaS_2 + GeSe_2$ *system (at.% Se is indicated). (b) Dependence of the emission intensity at the maximum* (λ_m *1180 nm) on the content of Se (excitation wavelength 532 nm, temperature 80 K).*

gap of chalcogenide glass semiconductors. Optical transitions of electrons into this band can cause luminescence with the energy of light quanta close to the half-width of the band gap.

The glassy alloy $Ag_{0.05}Ga_{0.05}Ge_{0.95}S_2$ is characterized by the largest transparency window [8] among other glasses of the $AgGaSe_2 + GeS_2 \Leftrightarrow AgGaS_2 + GeSe_2$ system. This alloy was doped with an Er_2S_3 admixture. PL spectra investigation (**Figure 4**) was performed on the glasses $(100 - X)Ag_{0.05}Ga_{0.05}Ge_{0.95}S_2$-(X) Er_2S_3, where X = 0.42, 0.25, and 0.18 mol.% (0.27, 0.16, and 0.12 at.% Er, respectively).

The excitation was performed by a laser with 980-nm wavelength which corresponds to the $^4I_{15/2} \rightarrow {}^4I_{11/2}$ transition, whereas the PL emission band is due to the $^4I_{13/2} \rightarrow {}^4I_{15/2}$ transition in Er^{3+} ion, respectively. The PL intensity increases with Er content. The position of the luminescence maximum at 1540 nm does not depend on the content of Er or other components of the glass-forming matrix. The effective width $\Delta\lambda_{eff}$ of the spectra PL was calculated for these glasses according to the formula

$$\Delta\lambda_{eff} = \frac{\int I(\lambda)d\lambda}{I_{max}} \tag{1}$$

where $I(\lambda)$ is the emission intensity at wavelength λ and I_{max} is the maximum emission intensity.

The maximum $\Delta\lambda_{eff}$ value was found for the sample with 0.27 at.% Er (61 nm); this decreases the alloys with 0.16 at.% (52 nm) and 0.112 at.% Er (52 nm). The widening of the PL band in erbium-doped glasses was associated by the authors of [10] with the formation of clusters in the samples where Ga/Er < 10 is performed. Among the glassy alloys investigated here, the PL band widens in the sample with 0.27 at.% Er, and the above inequality is fulfilled. The microstructure study of alloys [11] showed the existence of inhomogeneities of 6–7 μm size the concentration of which increases with the amount of erbium. The generated inhomogeneities are

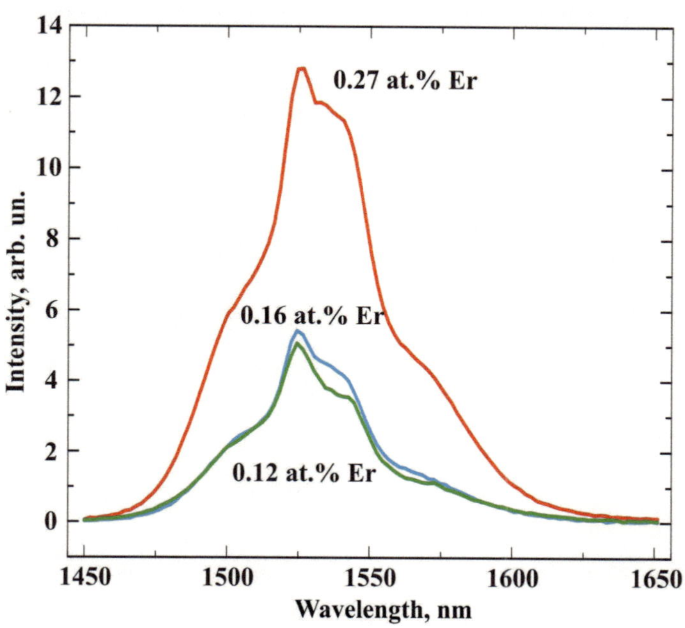

Figure 4.
Conversion FL spectra of the $Ag_{0.05}Ga_{0.05}Ge_{0.95}S_2$-$Er_2S_3$ glasses.

good drains of defects which serve as the basis for cluster formation. Higher amounts of erbium favor the formation of clusters which include it. Thus, the PL intensity in the glasses is due to the emission of erbium ions which are uniformly distributed in the glass-forming matrix, as well as those which are located near the inhomogeneities and are involved in the formation of clusters. The study of static magnetization in the $Ag_{0.05}Ga_{0.05}Ge_{0.95}S_2$-$Er_2S_3$ glasses [12] confirmed the formation of clusters. The calculated number of erbium ions in the cluster was estimated there as 1–1.5 × 10^3.

The excitation at 980 nm yielded, in addition to the conversion PL (maximum at 1540 nm, **Figure 4**), the up-conversion PL bands in the visible and near-infrared spectral range at room temperature (**Figures 5** and **6**).

For all glasses, the luminescence is represented by three maxima at 520, 657, and 855 nm, which correspond to the radiative transitions $^2H_{11/2} \rightarrow {}^4I_{15/2}$, $^4F_{9/2} \rightarrow {}^4I_{15/2}$, and $^4S_{3/2} \rightarrow {}^4I_{13/2}$ in Er^{3+} ions, respectively. In addition, a wide PL band in the range of 695–810 nm with an emission maximum at 765 nm was detected for the sample containing 0.27 at.% Er that cannot be interpreted by any radiative transition in erbium ion.

The intensity of the up-conversion PL bands depends on the laser excitation power (I_{IR}) which is expressed by the formula $I_{PL} \propto I^n_{IR}$, where n is the number of infrared photons per one PL photon. The number (n) can be found by the slope of the dependence of log (I_{PL}) on log (I_{IR}). PL spectral dependences for the sample with 0.27 at.% Er at different laser excitation powers are shown in **Figure 6**; these are also typical of the main maxima (520, 657, and 855 nm) in samples with less erbium content. The logarithmic dependence of the PL intensity on the excitation power [13], log (I_{PL}) from log (P_{IR}), is plotted in **Figure 7**.

Studies show that two photons with a wavelength of 980 nm are needed for the emission of one I_{PL} photon. An energy level diagram in erbium ions when excited by $h\nu_{980}$ quanta is shown in **Figure 8**. The absorption by the ground-state erbium ion of one $h\nu_{980}$ photon excites it to $^4I_{11/2}$ state. Such a transition can also take place via energy transfer (ET) from the neighboring excited erbium ion.

The excited state $^4F_{7/2}$ may be attained by the successive absorption of two $h\nu_{980}$ photons as well as by energy transfer to Er^{3+} ion in the $^4I_{11/2}$ state from another neighboring excited Er^{3+} ion [14]:

$$^4I_{15/2} + h\nu_{980} \rightarrow {}^4I_{11/2} + h\nu_{980} \rightarrow {}^4F_{7/2} \quad (2)$$

$$^4I_{11/2} + {}^4I_{11/2} \rightarrow {}^4I_{15/2} + {}^4F_{7/2} \quad (3)$$

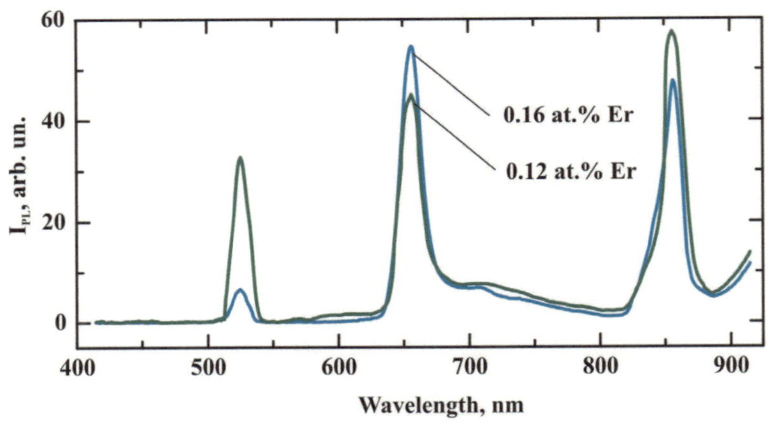

Figure 5.
Up-conversion FL spectra of the $Ag_{0.05}Ga_{0.05}Ge_{0.95}S_2$-$Er_2S_3$ glasses.

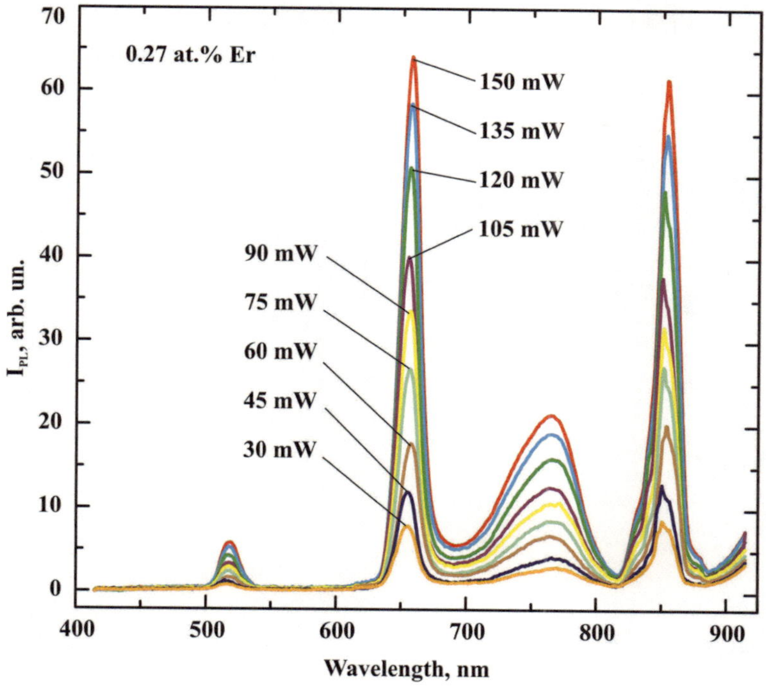

Figure 6.
Up-conversion FL spectra of the $Ag_{0.05}Ga_{0.05}Ge_{0.95}S_2$-$Er_2S_3$ glasses for various excitation power.

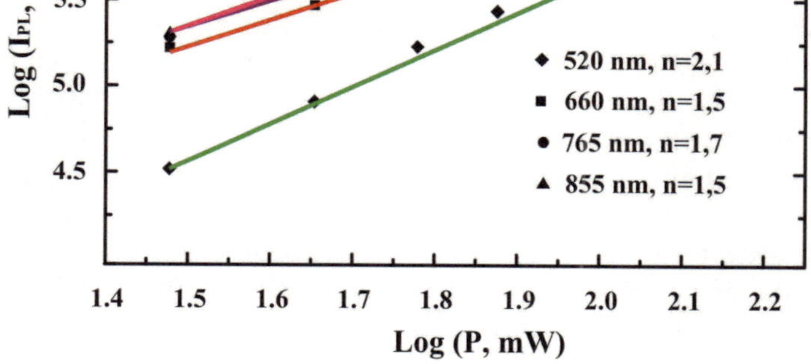

Figure 7.
Dependence of the luminescence intensity on the laser excitation power for the glass with 0.27 at.% Er.

The states $^2H_{11/2}$ and $^4S_{3/2}$ arise from non-radiative relaxation of the state $^4F_{7/2}$ due to the small energy distance between them. The non-radiative relaxation of Er^{3+} from the state $^4S_{3/2}$ to $^4F_{9/2}$ is unlikely due to large energy gap between them (about 3000 cm^{-1}) and low energy of phonons. According to Raman spectroscopy study, the energy of phonons for these glasses is close to 300–400 cm^{-1} [13]. Consequently, the state $^4F_{9/2}$ may arise by the absorption of the $h\nu_{980}$ photon

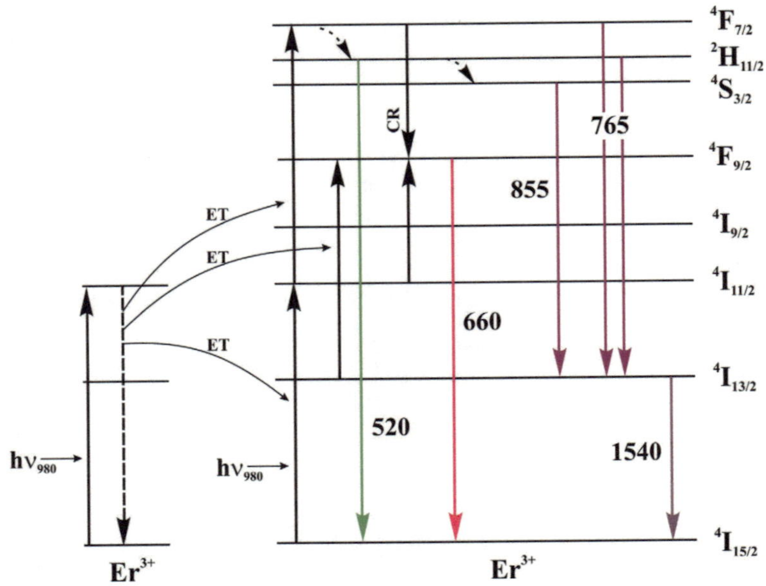

Figure 8.
Energy level diagram of erbium ions (for the $Ag_{0.05}Ga_{0.05}Ge_{0.95}S_2$-$Er_2S_3$ glasses, excitation by 980 nm).

(transition $^4I_{13/2} \rightarrow {}^4F_{9/2}$) or through the ET from the neighboring excited erbium ion. Additionally, the state $^4F_{9/2}$ state may appear due to cross-relaxation (CR):

$$^4F_{7/2} + {}^4I_{11/2} \rightarrow 2^4F_{9/2} \tag{4}$$

Such a mechanism of occurrence of the excited state $^4F_{9/2}$ would explain the decrease in the intensity of the green (520 nm) and the amplification of the red PL band with an increase in the content of Er^{3+} ions when the distance between them decreases and the conditions favor ET.

The increase of PL intensity in the chalcogenide glasses may be due to the increase in erbium content and the change in the composition of the glass-forming matrix. It should be noted that an increase in erbium content in glasses can lead to the concentration quenching of PL [15] as well as to the crystallization of the host. Therefore, increasing the PL efficiency requires the selection of optimal composition of the glass-forming matrix and adding into its composition the maximum concentration of erbium that will not cause PL quenching and the crystallization of glass. The choice of the components of the glass-forming matrix should take into account their glass-forming ability and the ability of sustaining large amount of REs and also maintaining a wide transparency window in the visible and infrared range when adding admixtures. With the above considerations, the Ga_2S_3-La_2S_3-Er_2S_3 system is of interest, in our opinion. Using the technique in paragrap. 2, we successfully introduced up to 40 mol.% La_2S_3 and 3 mol.% Er_2S_3 into the glass. The shift of the optical absorption edge does not exceed 0.13 eV with Er_2S_3 doping and 0.10 eV with the addition of 30–40 mol.% La_2S_3 [16]. Thus, the luminescence properties of the Ga_2S_3-La_2S_3-Er_2S_3 system glasses at room temperature and 80 K were investigated with a considerable variation in the host composition. The component composition of the glasses is listed in **Table 1**.

Photoluminescence spectra in the 480–1700 nm range upon laser excitation at 488 nm wavelength at 80 K are shown in **Figures 9** and **10**. The intense maxima at 550, 855, 985, 1100, and 1540 nm correspond to 4f intra-shell transitions $^4S_{3/2} \rightarrow {}^4I_{15/2}$, $^4S_{3/2} \rightarrow {}^4I_{13/2}$, $^4I_{11/2} \rightarrow {}^4I_{15/2}$, $^2H_{11/2} \rightarrow {}^4I_{11/2}$, $^4I_{13/2} \rightarrow {}^4I_{15/2}$ in Er^{3+} ions.

Sample no.	Ga_2S_3	La_2S_3	Er_2S_3
		(mol. %)	
1	64	35	1
2	62	35	3
3	59	40	1
4	57	40	3

Table 1.
Component composition of the Ga_2S_3-La_2S_3-Er_2S_3 glasses.

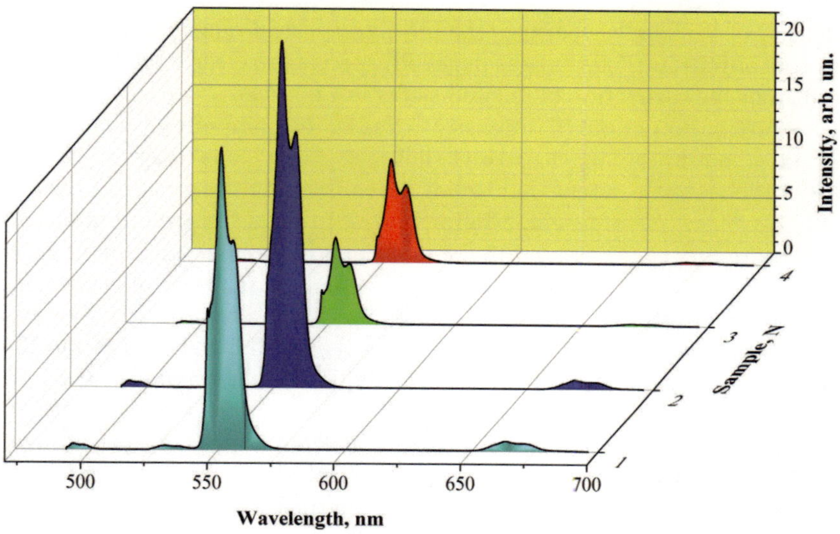

Figure 9.
PL spectra of the Ga_2S_3-La_2S_3-r_2S_3 glasses at 80 K (visible range).

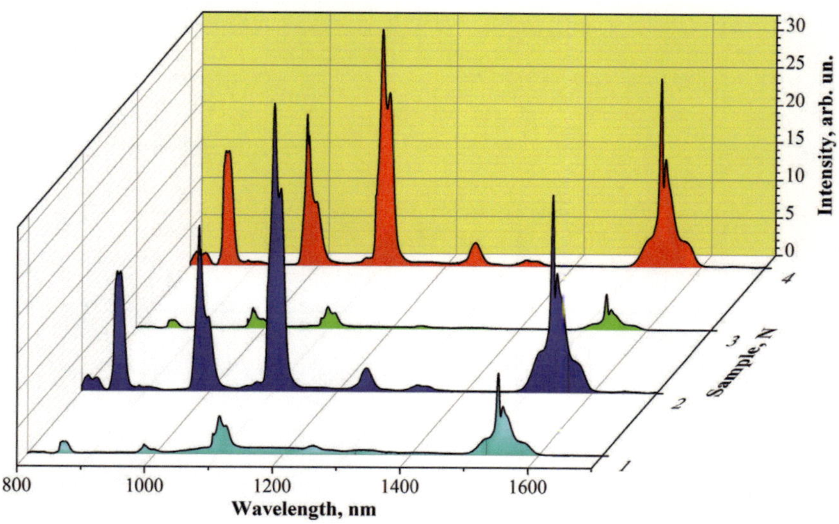

Figure 10.
PL spectra of the Ga_2S_3-La_2S_3-Er_2S_3 glasses at 80 K (infrared range).

The lower-intensity maxima at 492, 660, 810, and 1245 nm correspond to the transitions $^4F_{7/2} \rightarrow {}^4I_{15/2}$, $^4F_{9/2} \rightarrow {}^4I_{15/2}$, $^4I_{9/2} \rightarrow {}^4I_{15/2}$, $^4F_{7/2} \rightarrow {}^4I_{9/2}$ in erbium ions, respectively.

PL intensity increases in the visible and infrared ranges with Er_2S_3 concentration at constant content of La_2S_3. The best medium for PL (the most intense PL) is the glass-forming matrix with 35 mol.% La_2S_3 and 3 mol.% Er_2S_3. PL intensity decreases when La_2S_3 content increased to 40 mol.%. A similar effect was observed by Ivanova in [17] in the glasses of the GeS_2-Ga_2S_3 system, explaining this by an increase in the number of homopolar metallic bonds. It is likely that the number of La-La-type bonds increases in the Ga_2S_3-La_2S_3-Er_2S_3 glasses with increasing La_2S_3 content.

Changing the component composition of glass leads to redistribution of the intensity of PL bands. The maximum at 550 nm is dominating in the visible range, and its intensity decreases with increasing La_2S_3 content. The greatest increase in the intensity of PL bands occurs for the infrared spectral range with an increase in erbium concentration. The changes in the PL spectra are clearly related to the mechanism of the realization of excited states in Er^{3+} ions.

The redistribution of the PL intensity is due to the competing ways of achieving excited states in erbium ions. **Figure 11** shows an energy level diagram of erbium ions [18] which reveals the mechanism of PL emission in the glasses of the Ga_2S_3-La_2S_3-Er_2S_3 system. Accordingly, erbium ions jump from the ground to the excited state of $^4F_{7/2}$ upon excitation by 488 nm wavelength. Intense emission of green and infrared light (550, 1100, and 855 nm) results when the majority of erbium ions non-radiatively relax to the states $^2H_{11/2}$ and $^4S_{3/2}$. Only a small fraction of erbium undergoes transitions $^4F_{7/2} \rightarrow {}^4I_{15/2}$ and $^4F_{7/2} \rightarrow {}^4I_{9/2}$ as PL bands at 492 and 1245 nm are characterized by low intensity. Additionally, PL with a maximum at 660 nm occurs due to the cross-relaxation process CR resulting in the excited state $^4F_{9/2}$.

PL spectra of the Ga_2S_3-La_2S_3-Er_2S_3 glasses at room temperature are shown in **Figures 12** and **13**. Compared to the spectra at 80 K, the PL intensity at 530 nm increased slightly (transition $^2H_{11/2} \rightarrow {}^4I_{15/2}$), and the band intensity at 660 nm increased (transition $^4F_{9/2} \rightarrow {}^4I_{15/2}$). Additionally, the intensity of the maxima of

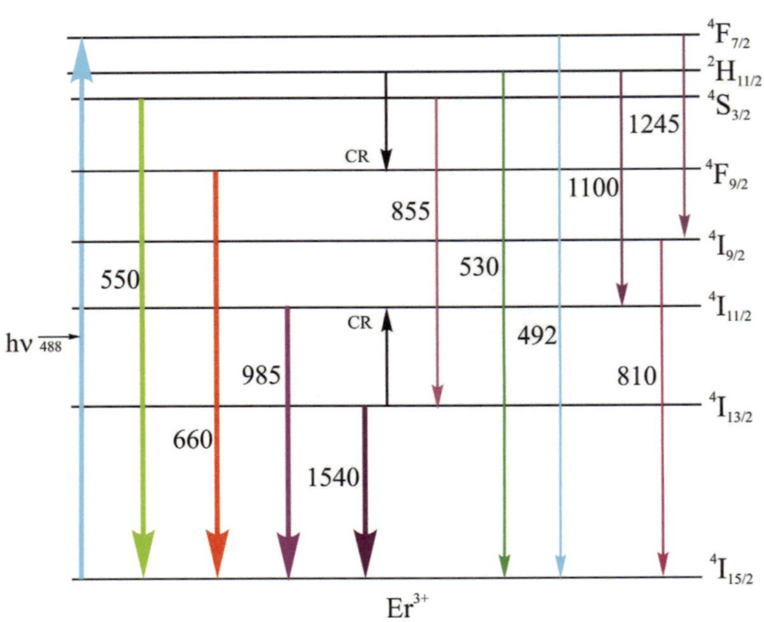

Figure 11.
Energy level diagram of erbium ions (for the Ga_2S_3-La_2S_3-Er_2S_3 glasses).

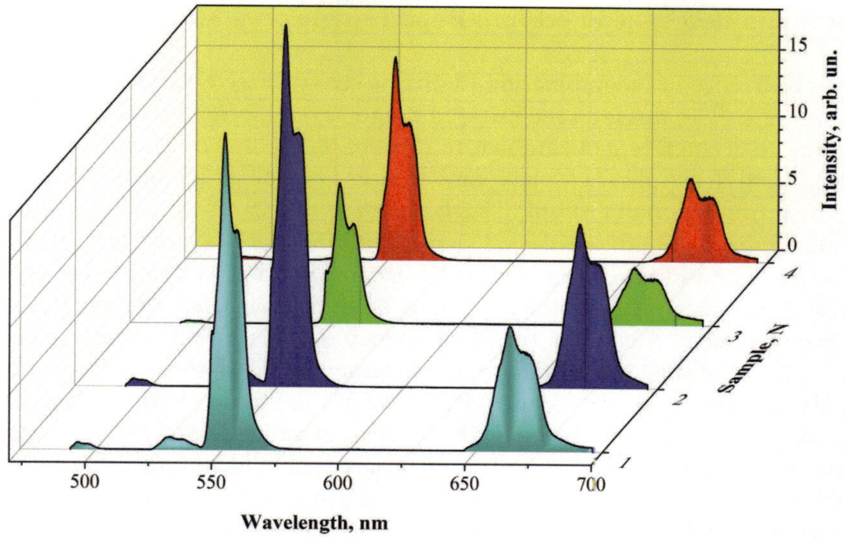

Figure 12.
PL spectra of the Ga_2S_3-La_2S_3-Er_2S_3 glasses at room temperature (visible range).

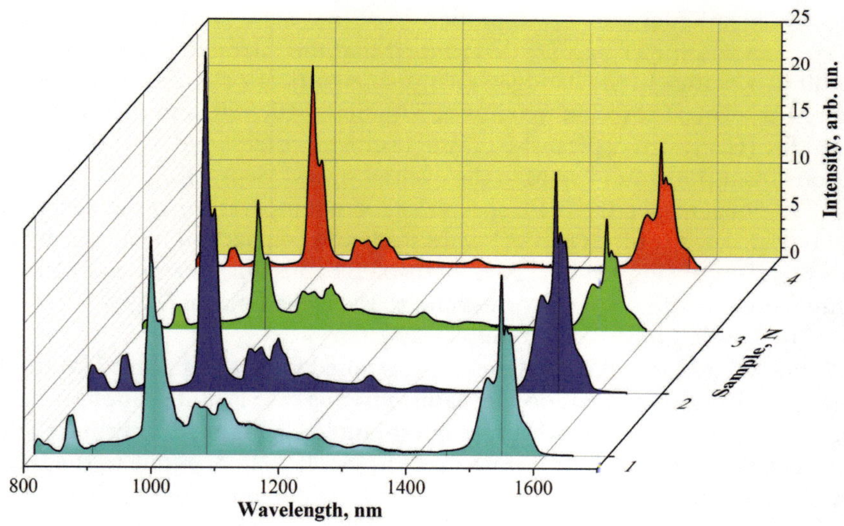

Figure 13.
PL spectra of the Ga_2S_3-La_2S_3-Er_2S_3 glasses at room temperature (infrared range).

1100 and 855 nm significantly decreased, and a weak PL band at 1050 nm was observed in the infrared spectral range. Such intensity redistribution of the PL maxima is associated with a change in the concentration of erbium ions in different excited states. This is related to the fact that when the temperature rises, the phonon subsystem of the glass-forming matrix is transformed, so the probability of the energy exchange of the neighboring erbium ions increases. As a result, the role of cross-relaxation processes increases and the number of erbium ions in the state $^4F_{9/2}$ increases. Thus the intensity of the PL band with a maximum at 660 nm increases. Additionally, CR affects the decrease of the number of erbium ions in the states $^2H_{11/2}$ and $^4S_{3/2}$ leading to the decrease of the intensity of the maxima at 1100 and 855 nm at room temperature. The band with a maximum at 1050 nm (transition $^4F_{3/2} \rightarrow {}^4I_{9/2}$) is completely overlapped by the intense PL maximum at 1100 nm at 80 K.

4. Photoluminescence in erbium-doped single crystals

The occurrence of recombination PL that is characterized by wide emission bands in the visible, rarely in the infrared range, is quite often found in single-crystalline chalcogenide semiconductors at room temperature or low temperatures [19, 20]. From the application viewpoint of laser technology and telecommunications, the introduction of erbium to such materials has significant advantages since the emission bands of erbium-doped glasses and single crystals are usually intense and narrow. The addition of erbium is also accompanied by the extinction of PL radiation [21] which is due to the crystalline or amorphous chalcogenide host. The influence of external factors on the PL efficiency in erbium-doped chalcogenides is limited due to the shielding of radiative transitions in the 4f-shell of Er by outer shells. At the same time, the local environment of erbium ion, both in glasses and single crystals, affects the efficiency of PL emission which is why it is extremely important for the design of fluorescent materials. It should be noted that the single crystals, unlike glasses, do not permit a wide variation of the component composition since it is limited by the solid solubility of impurities and the homogeneity of the crystalline compound. It is more difficult to select the composition of the host to obtain effective emission in erbium-doped single crystals; therefore, PL is less common in these than in the corresponding amorphous media.

The investigation of the optical properties of the single crystals $(Ga_{55}In_{45})_2S_{300}$ and $(Ga_{54.59}In_{44.66}Er_{0.75})_2S_{300}$ [7] determined that the introduction of erbium does not result in a change in the bandgap energy or significant changes in the electronic structure [22]. The absorption maxima at 530, 660, 810, 980, and 1530 nm [7] recorded for $(Ga_{54.59}In_{44.66}Er_{0.75})_2S_{300}$ single crystal correspond to 4f intra-shell transitions from the ground state to the excited states $^2H_{11/2}$, $^4F_{9/2}$, $^4I_{9/2}$, $^4I_{11/2}$, and $^4I_{13/2}$ in Er^{3+} ions, respectively. PL spectra at room temperature in the visible and close infrared range (600–1650 nm) were studied when excited by a laser with 532 nm wavelength (**Figure 14**). Two intense PL bands with maxima at 810 and 1540 nm were recorded, which correspond to the transitions of $^4I_{9/2} \rightarrow {}^4I_{15/2}$ and $^4I_{13/2} \rightarrow {}^4I_{15/2}$ in Er^{3+} ions, respectively.

Er^{3+} ions excited by a 532 nm laser are promoted from the ground $^4I_{15/2}$ to the excited state $^2H_{11/2}$. The emission mechanism in this case is quite simple as demonstrated at an energy transition diagram in erbium ions (**Figure 15**). Erbium ions cannot relax non-radiatively from the state $^4S_{3/2}$ to $^4F_{9/2}$ or $^4I_{9/2}$ due to the large energy distance and the low phonons energy (200–300 cm^{-1} for the single crystals $(Ga_{55}In_{45})_2S_{300}$ and $(Ga_{54.59}In_{44.66}Er_{0.75})_2S_{300}$ [23]). Therefore, the appearance of

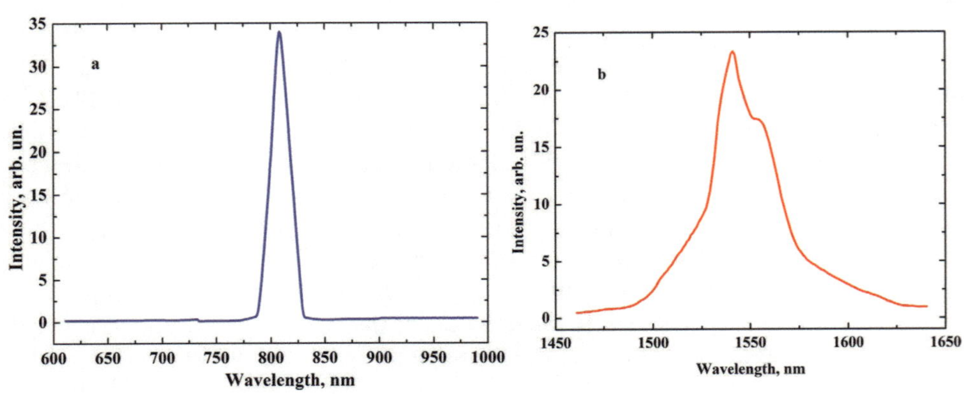

Figure 14.
Visible (a) and NIR (b) PL spectra of the $(Ga_{54.59}In_{44.66}Er_{0.75})_2S_{300}$ single crystal (excitation by 532 nm).

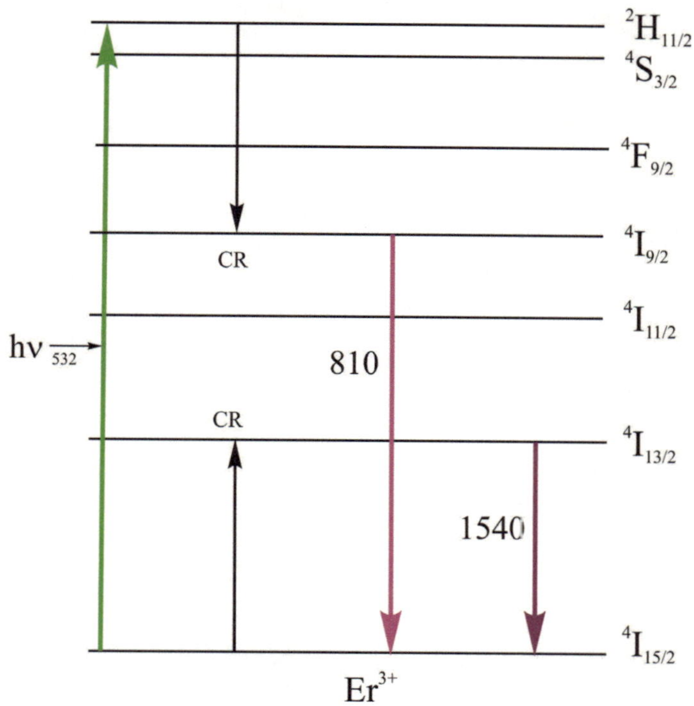

Figure 15.
Energy level diagram in erbium ions (for a $(Ga_{54.59}In_{44.66}Er_{0.75})_2S_{300}$ single crystal, excitation by 532 nm).

excited states $^4I_{9/2}$ involves the cross-relaxation process CR. Energy transfer between adjacent erbium ions in the states $^2H_{11/2}$ and $^4I_{15/2}$ promotes one ion into the excited state $^4I_{9/2}$ and the other into $^4I_{13/2}$:

$$^2H_{11/2} + {}^4I_{15/2} \rightarrow {}^4I_{9/2} + {}^4I_{13/2} \quad (5)$$

Such a mechanism yields substantial numbers of erbium ions in excited states $^4I_{9/2}$ and $^4I_{13/2}$, which results in PL with maxima at 810 and 1540 nm.

Similar to glasses, we also investigated single crystals in which two rare earth metals were introduced. $(Ga_{55}In_{45})_2S_{300}$ and $(Ga_{54.59}In_{44.66}Er_{0.75})_2S_{300}$ single crystals were obtained by a solution-melt method [24] selected in accordance with the phase diagram of the Ga_2S_3-La_2S_3 system. The bandgap energy of the single crystals little changes upon erbium doping at 2.01 and 1.99 eV for $(Ga_{55}In_{45})_2S_{300}$ and $(Ga_{54.59}In_{44.66}Er_{0.75})_2S_{300}$, respectively [24]. PL of $(Ga_{54.59}In_{44.66}Er_{0.75})_2S_{300}$ single crystal in the 500–1700 nm range was investigated at room temperature (**Figure 16**). Four intense PL bands with maxima at 525, 545, 980, and 1540 nm and one low-intensity band at 660 nm were recorded. They correspond to the transitions in Er^{3+} ions from excited states $^2H_{11/2}$, $^4S_{3/2}$, $^4I_{11/2}$, $^4I_{13/2}$, and $^4F_{9/2}$ to the ground state.

Energy transition diagram (**Figure 17**) shows that the wavelength of 810 nm can excite erbium ions from the ground state $^4I_{15/2}$ to the state $^4I_{9/2}$. Er^{3+} ions, either by directly absorbing light quanta or through ET, can be promoted from the state $^4I_{13/2}$ to $^2H_{11/2}$ resulting in PL at 525 nm. Additionally, after non-radiative relaxation of erbium ions to the state $^4S_{3/2}$, there is a PL band with a maximum at 545 nm [25]. The appearance of excited states $^4F_{9/2}$ and $^4I_{11/2}$ occurs by cross-relaxation CR 2:

$$^2H_{11/2} + {}^4I_{13/2} \rightarrow {}^4F_{9/2} + {}^4I_{11/2} \quad (6)$$

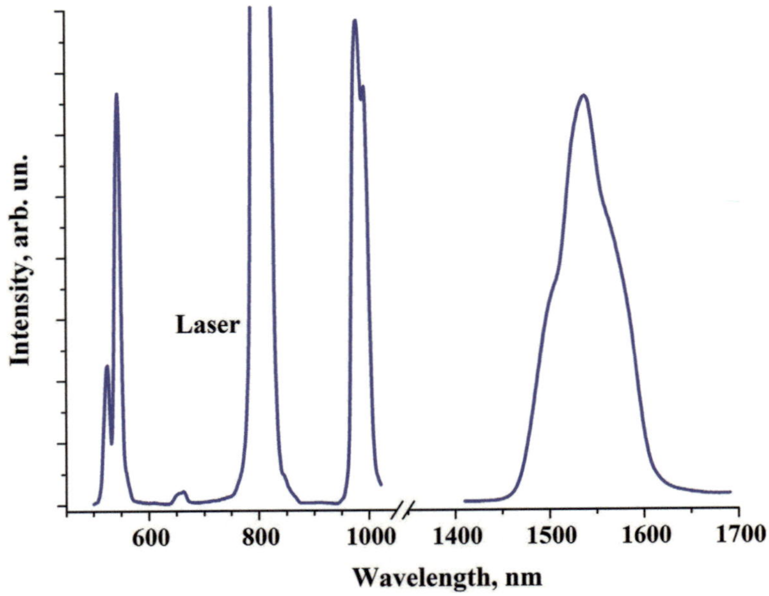

Figure 16.
PL of the $(Ga_{69.75}La_{29.75}Er_{0.5})_2S_{300}$ single crystal (excitation by 810 nm).

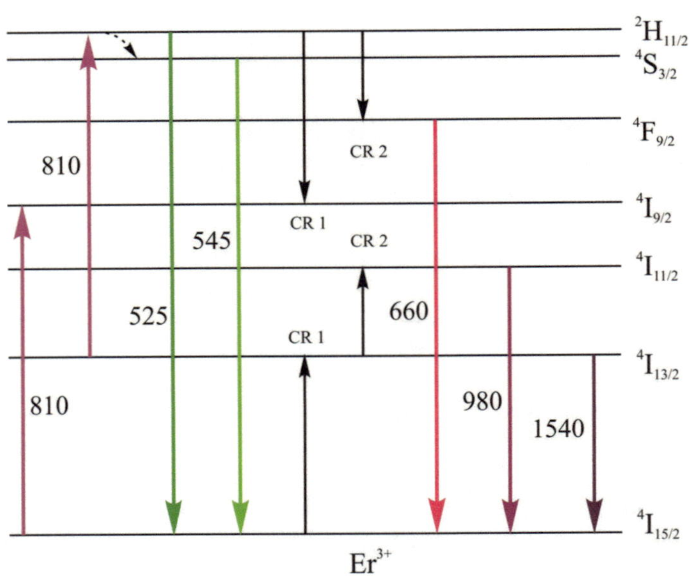

Figure 17.
Energy level diagram in erbium ions (for a $(Ga_{69.75}La_{29.75}Er_{0.5})_2S_{300}$ single crystal, excitation by 810 nm).

The transition from these states to the ground state produces PL at 660 and 980 nm. An intense PL band at 1540 nm wavelength occurs due to the formation of a significant number of Er^{3+} ions in the state $^4I_{13/2}$ by cross-relaxation CR 1:

$$^2H_{11/2} + {}^4I_{15/2} \rightarrow {}^4I_{9/2} + {}^4I_{13/2} \quad (7)$$

The excitation of the single crystal yields also an intense up-converted green PL (**Figure 16**). Therefore, $(Ga_{54.59}In_{44.66}Er_{0.75})_2S_{300}$ single crystal can be recommended as a material for the manufacture of light converters.

5. The influence of temperature on the photoluminescent properties of glasses and crystals

Doping the binary and ternary chalcogenides, particularly by rare earth metals, creates materials for the design of active and passive media for laser technology, photonic devices, light converters, and telecommunications. Special attention is given to the crystalline and amorphous environments that can exhibit high-intensity PL under the influence of external factors.

Additionally, modern optoelectronic industry actively researches and implements high-precision temperature sensors based on the sensitivity of PL emission to temperature changes. The design of such devices requires the investigation of the effect of temperature on the mechanism of the achievement of excited states and the processes of the relaxation of erbium ions when the PL intensity changes in various temperature ranges. This will allow optimizing the component composition of crystals and glasses to produce effective luminescent materials for the development of non-contact optical sensors.

PL of the $Ag_{0.05}Ga_{0.05}Ge_{0.95}S_2$-$Er_2S_3$ glasses in 600–1050 nm range was investigated in a temperature range of 80–300 K upon laser excitation at 532 nm [26]. Recorded PL bands with maxima at 660, 860, and 980 nm correspond to the radiative transitions $^4F_{9/2} \rightarrow {}^4I_{15/2}$, $^4S_{3/2} \rightarrow {}^4I_{13/2}$, $^4I_{11/2} \rightarrow {}^4I_{15/2}$ in erbium ions, respectively. The dependences of PL intensity on wavelength at different temperatures for the sample with 0.27 at.% Er are plotted in **Figure 18** (also typical of the glasses with 0.12, 0.16 at.% Er). The temperature increase leads to changes in the ratio of PL intensities, with the intensity of all bands decreasing at temperatures above 180 K. The PL band with a maximum at 860 nm is the most sensitive to temperature changes. The PL emission mechanism is well explained by the energy level diagram of Er^{3+} ions (**Figure 19**).

Illumination by 532 nm wavelength promotes erbium ions to the excited state $^2H_{11/2}$. The states $^4F_{9/2}$, $^4I_{9/2}$, and $^4I_{11/2}$ result in the cross-relaxation processes CR1, CR2, and CR3. Er^{3+} ions in the $^2H_{11/2}$ state can also nonradiatively relax to the $^4S_{3/2}$ state due to small energy gap between them. The integral intensity of the emission bands was calculated from the results of the investigation of PL spectra.

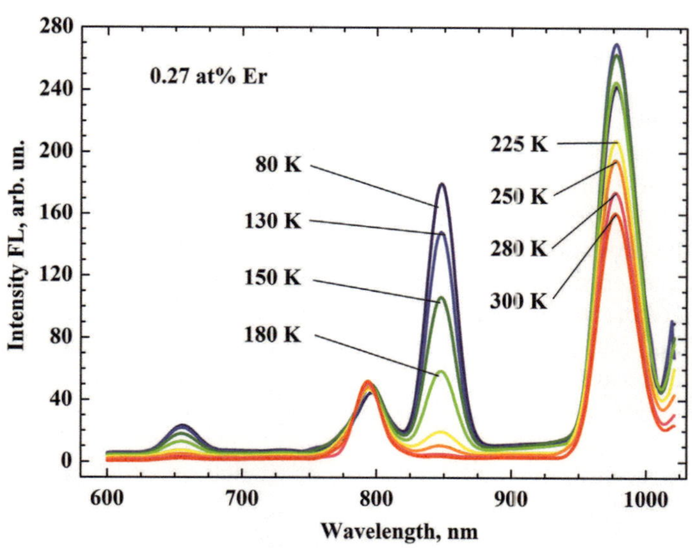

Figure 18.
PL spectra of the $Ag_{0.05}Ga_{0.05}Ge_{0.95}S_2$-$Er_2S_3$ glass (0.27 at.% Er) at various temperatures.

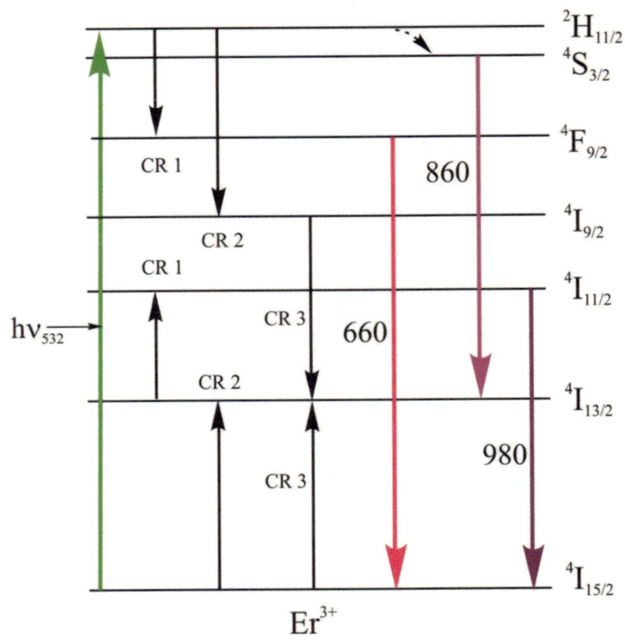

Figure 19.
Energy level diagram in erbium ions (for the $Ag_{0.05}Ga_{0.05}Ge_{0.95}S_2$-$Er_2S_3$ glasses, excitation by 532 nm).

The dependence of integral PL intensity on temperature is shown in **Figure 20** for the glass with 0.27 at.% Er.

The dependence of PL intensity on temperature can be described using the probability of radiative and non-radiative processes by the formula [27]:

$$I(T) = \frac{I_0}{1 + \frac{\omega_{nr}}{\omega_r}\exp\left(-\frac{E_t}{kT}\right)} \tag{8}$$

where I_o is a constant, k is Boltzmann's constant, ω_{nr}, ω_r is the probability of non-radiative and radiative processes, respectively, and E_t is the thermal activation energy of luminescence.

The I(T) dependence (**Figure 20**, solid line) was calculated from Eq. (8) for the band with the 860 nm maximum, and the corresponding activation energy E_t was calculated as 90 ± 6 meV [26]. According to the diagram in **Figure 19**, this energy determines the activation of erbium ions from the state $^4S_{3/2}$ в $^2H_{11/2}$ (i.e., the energy gap between these states). It should be noted that changes in PL intensity is complex. Therefore we calculated the logarithm of the ratio of integral PL intensities ln (I_{980}/I_{660}) and plotted its dependence on temperature (**Figure 21**). This dependence is linear in the range of 125–300 K. The sensitivity of the sample as the temperature sensor was calculated according to the results of these studies as $0.43\ K^{-1}$.

Erbium-doped single crystals may also exhibit changes of PL intensity with temperature. PL spectra of the single crystal $(Ga_{54.59}In_{44.66}Er_{0.75})_2S_{300}$ were investigated at 150, 200, 250, and 300 K upon laser excitation at 980 nm wavelength (**Figures 22** and **23**). The position and shape of the maxima do not change but the intensity of PL increases with temperature. For the majority of semiconductor

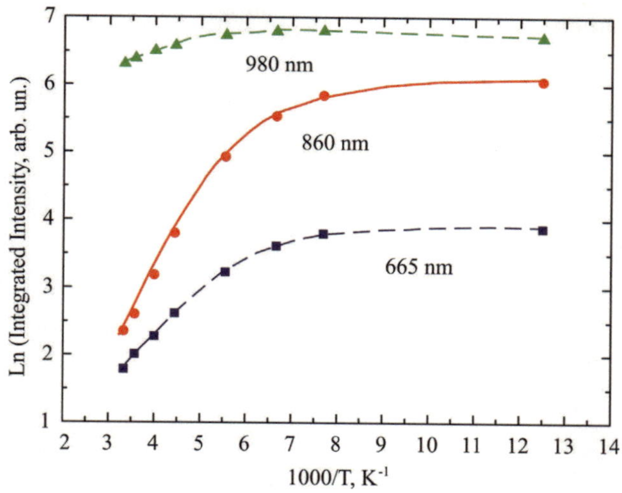

Figure 20.
Temperature dependence of integral PL intensity of the $Ag_{0.05}Ga_{0.05}Ge_{0.95}S_2$-$Er_2S_3$ *glass (0.27 at.% Er).*

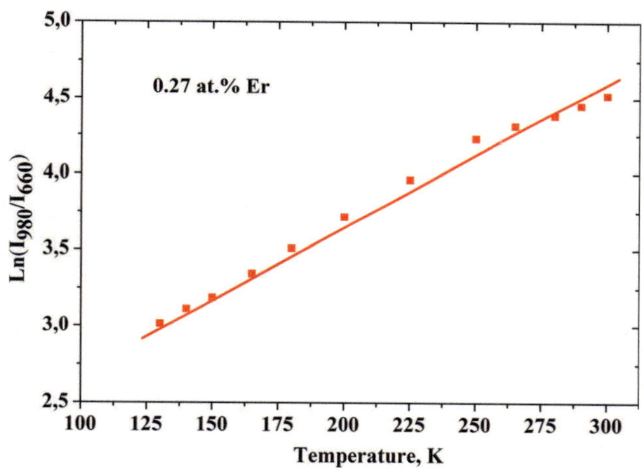

Figure 21.
Ratio of integral PL intensity of the glasses of the $Ag_{0.05}Ga_{0.05}Ge_{0.95}S_2$-$Er_2S_3$ *system (0.27 at.% Er).*

materials where the recombination luminescence was recorded [28], the PL intensity decreases with increasing temperature. At the same time, the emission intensity may increase with temperature in crystals and glasses where PL is associated with transitions in the 4f-shell of erbium ions. This is due to the fact that the neighboring erbium ions are in different excited states. As the temperature increases, the phonon subsystem of the crystal changes which contributes to the cross-relaxation processes of Er^{3+} ions. If a crystal is excited by 980 nm wavelength, the absorption of two photons promotes erbium ion into the state $^4F_{7/2}$ (**Figure 22**). Due to small energy gap, Er^{3+} ions can non-radiatively relax to the state $^2H_{11/2}$. Excited states $^4I_{13/2}$ and $^4I_{9/2}$ result from the CR process:

$$^4I_{15/2} + {}^2H_{11/2} \rightarrow {}^4I_{13/2} + {}^4I_{9/2} \tag{9}$$

Therefore, CR processes play greater role with increasing temperature, which will contribute to a higher concentration of erbium ions in the $^4I_{9/2}$ and $^4I_{13/2}$ states and the increase in PL intensity.

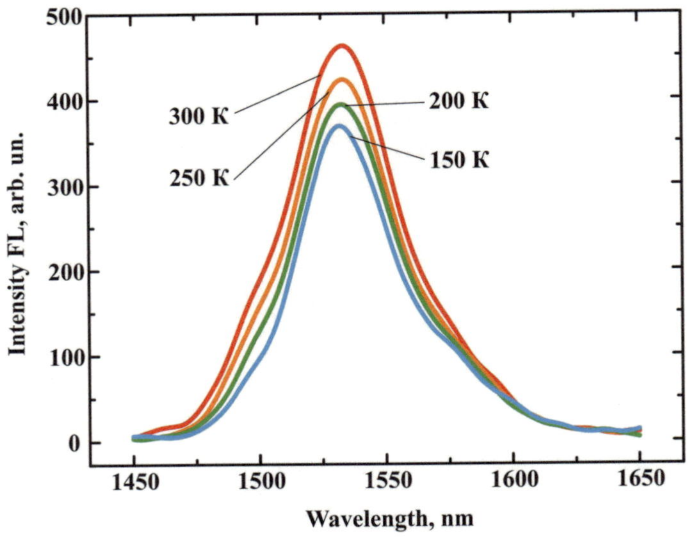

Figure 22.
Conversion PL spectra of the $(Ga_{54.59}In_{44.66}Er_{0.75})_2S_{300}$ *single crystal at various temperatures.*

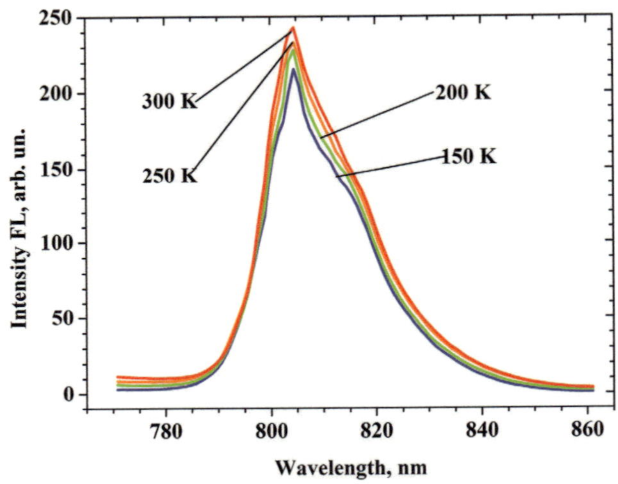

Figure 23.
Up-conversion PL spectra of the $(Ga_{54.59}In_{44.66}Er_{0.75})_2S_{300}$ *single crystal at various temperatures.*

The integral PL intensity for both maxima was calculated from the spectra of up-conversion and conversion PL of $(Ga_{54.59}In_{44.66}Er_{0.75})_2S_{300}$ single crystal (**Figure 24**). The plots of integral PL intensity linearly depend on temperature. From the results of the temperature dependence of the integral PL intensity, the sensitivity was calculated as $1.187 \times 10^{-3}\ K^{-1}$ for PL at 805 nm and $1.818 \times 10^{-3}\ K^{-1}$ for the maximum at 1540 nm. It should be noted that the sensitivity defined as the ratio of the integral intensities of the two PL bands is higher in the $Ag_{0.05}Ga_{0.05}Ge_{0.95}S_2$-$Er_2S_3$ glasses. The sensitivity in $(Ga_{54.59}In_{44.66}Er_{0.75})_2S_{300}$ single crystal was calculated separately for each PL band (maxima at 805 and 1540 nm), the latter being in the operating range of fiber-optic networks and telecommunication devices.

Therefore, in a limited temperature range, the $Ag_{0.05}Ga_{0.05}Ge_{0.95}S_2$-$Er_2S_3$ glasses (with 0.27 at.% Er) and the single crystals $(Ga_{54.59}In_{44.66}Er_{0.75})_2S_{300}$ can be used to design non-contact optical temperature sensors.

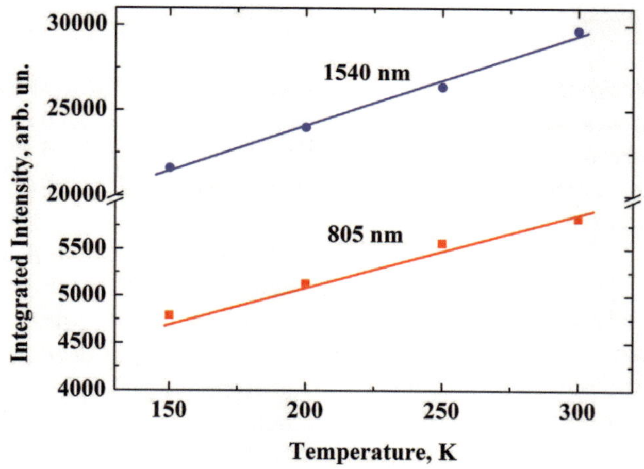

Figure 24.
Temperature dependence of integral PL intensity of the $(Ga_{54.59}In_{44.66}Er_{0.75})_2S_{300}$ *single crystal.*

6. Conclusions

The principal mechanisms of the occurrence of PL due to the transitions in the 4f-shell of erbium ions are presented on the basis of sulfide glasses and single crystals. The appearance of many emission bands in chalcogenide glasses (unlike single crystals) is due to the fact that in an amorphous medium, erbium ions may occupy several different positions in the glass-forming matrix. The change of the excitation wavelength leads to a change in the mechanism for the excited states in Er^{3+} ions and the emergence of some radiation bands and the extinction of others. Energy exchange processes (energy transfer, cross-relaxation) between the neighboring Er^{3+} ions strongly influence the intensity of the conversion and up-conversion PL. The investigation of the temperature dependence of PL indicates that erbium-doped chalcogenide semiconductors can be recommended as materials for the design of non-contact optical thermometers.

Author details

Volodymyr V. Halyan[1]* and Inna A. Ivashchenko[2]

1 Department of Experimental Physics and Technologies for Information Measuring, Lesya Ukrainka Eastern European University, Lutsk, Ukraine

2 Department of Inorganic and Physical Chemistry, Lesya Ukrainka Eastern European University, Lutsk, Ukraine

*Address all correspondence to: halyan.volodimir@eenu.edu.ua

References

[1] Frumar M, Wagner T. Ag doped chalcogenide glasses and their applications. Current Opinion in Solid State and Materials Science. 2003;7: 117-126. DOI: 10.1016/S1359-0286(03) 00044-5

[2] Kevshin A, Halyan V, Davydyuk G, Parasyuk O, Mazurets I. Concentration dependence of the optical properties of glassy alloys in the HgS-Ga_2S_3-GeS_2 system. Glass Physics and Chemistry. 2010;**36**:27-32. DOI: 10.1134/ S1087659610010050

[3] Yang G, Xia F, He X, Wen J, Chen G. Effects of Pb on thermal stability and crystallization kinetics of GeS_2-Sb_2S_3-PbS glasses. International Journal of Applied Glass Science. 2016;7:337-344. DOI: 10.1111/ijag.12181

[4] Plotnichenko V, Philippovskiy D, Sokolov V, Sukhanov M, Velmuzhov A, Churbanov M, et al. Infrared luminescence in Bi-doped Ge-S and As-Ge-S chalcogenide glasses and fibers. Optical Materials Express. 2014;**4**:366-374. DOI: 10.1364/ OME.4.000366

[5] Bordovsky G, Marchenko A, Rabchanova T, Seregin P, Terukov E, Ali H. Study of platinum impurity atom state in vitreous arsenic selenide. Semiconductors. 2012;**46**:878-881. DOI: 10.1134/S1063782612070056

[6] Marchenko A, Rabchanova T, Seregin P, Zharkoy A, Bobokhuzhaev K. Structural chemical states of dopant atoms of platinum and gold in glass-like arsenic selenides. Glass Physics and Chemistry. 2016;**42**: 43-49. DOI: 10.1134/S1087659616 010077

[7] Ivashchenko I, Danyliuk I, Olekseyuk I, Pankevych V, Halyan V. Phase equilibria in the quasi-ternary system Ag_2S-Ga_2S_3-In_2S_3 and optical properties of $(Ga_{55}In_{45})_2S_{300}$, $(Ga_{54.59}In_{44.66}Er_{0.75})_2S_{300}$ single crystals. Journal of Solid State Chemistry. 2015;**227**:255-264. DOI: 10.1016/j.jssc.2015.04.006

[8] Halyan V, Kevshyn A, Ivashchenko I, Shevchuk M. Effect of the substitution of S for Se on the optical absorption spectra of the glassy alloys $Ag_{1.6}Ga_{1.6}Ge_{31.2}S_{61.6-x}Se_x$. Physics and Chemistry of Solid State. 2016;**17**: 342-345. DOI: 10.15330/pcss.17.3. 342-345

[9] Mott N, Davis E. Electronic Processes in Non-Crystalline Materials. Oxford: Clarendon-Press; 1971. 437 p. DOI: 10.1002/crat.19720070420

[10] Koughia K, Munzar M, Tonchev D, Haugen C, Decorby R, McMullin J, et al. Photoluminescence in Er-doped Ge-Ga-Se glasses. Journal of Luminescence. 2005;**112**:92-96. DOI: 10.1016/j. jlumin.2004.09.002

[11] Halyan V, Kevshyn A, Kogut Y, Davydyuk G, Shevchuk M, Kažukauskas V, et al. Photoluminescence in Er-doped $AgGaS_2$-GeS_2 glasses. Physica Status Solidi C. 2009;**6**:2810-2813. DOI: 10.1002/pssc.200982576

[12] Halyan V, Konchits A, Shanina B, Krasnovyd S, Lebed O, Kevshyn A, et al. EPR of γ-induced defects and their effects on the photoluminescence in the glasses of the $Ag_{0.05}Ga_{0.05}Ge_{0.95}S_2$-$Er_2S_3$ system. Radiation Physics and Chemistry. 2015;**115**:189-195. DOI: 10.1016/j.radphyschem.2015.06.019

[13] Halyan V, Strelchuk V, Yukhymchuk V, Kevshyn A, Davydyuk G, Shevchuk M, et al. Role of structural ordering on optical properties of the glasses $Ag_{0.05}Ga_{0.05}Ge_{0.95}S_2$-$Er_2S_3$. Physica B: Condensed Matter. 2013;**411**: 35-39. DOI: 10.1016/j.physb.2012. 10.042

[14] Halyan V, Kevshyn A, Davydyuk G, Shevchuk M. Mechanism of anti-stokes photoluminescence in $Ag_{0.05}Ga_{0.05}Ge_{0.95}S_2$-$Er_2S_3$ glassy alloys. Glass Physics and Chemistry. 2013;**39**:52-56. DOI: 10.1134/S1087659613010069

[15] Tver'yanovich Y. Concentration quenching of luminescence of rare-earth ions in chalcogenide glasses. Glass Physics and Chemistry. 2003;**29**: 166-168. DOI: 10.1023/A:1023407125519

[16] Halyan V, Kevshyn A, Ivashchenko I, Olekseuk I, Danyliuk I, Shavarova G. Optical absorption of chalcogenide glass Ga_2S_3-La_2S_3 doped with erbium. Physics and Chemistry of Solid State. 2017;**18**: 342-346. DOI: 10.15330/pcss.18.3.342-346

[17] Ivanova Z, Zavadil J, Rao K. Compositional trends in low-temperature photoluminescence of heavily Er-doped GeS_2-Ga_2S_3 glasses. Journal of Non-Crystalline Solids. 2011; **357**:2443-2446. DOI: 10.1016/j.jnoncrysol.2010.11.075

[18] Kityk I, Halyan V, Yukhymchuk V, Strelchuk V, Ivashchenko I, Zhydachevskii Y, et al. NIR and visible luminescence features of erbium doped Ga_2S_3-La_2S_3 glasses. Journal of Non-Crystalline Solids. 2018;**498**:380-385. DOI: 10.1016/j.jnoncrysol.2018.03.024

[19] Chen F, Jie W. Growth and photoluminescence properties of CdS solid solution semiconductor. Crystal Research and Technology. 2007;**42**: 1082-1086. DOI: 10.1002/crat.200710955

[20] Era K, Shionoya S, Washizawa Y. Mechanism of broad-band luminescences in ZnS phosphors—I. Spectrum shift during decay and with excitation intensity. Journal of Physics and Chemistry of Solids. 1968;**29**: 1827-1841. DOI: 10.1016/0022-3697(68)90167-4

[21] Kostka P, Zavadil J, Iovu M, Ivanova Z, Furniss D, Seddon A. Low-temperature photoluminescence in chalcogenide glasses doped with rare-earth ions. Journal of Alloys and Compounds. 2015;**648**:237-243. DOI: 10.1016/j.jallcom.2015.05.135

[22] Khyzhun O, Halyan V, Danyliuk I, Ivashchenko I. Electronic structure of $(Ga_{55}In_{45})_2S_{300}$ and $(Ga_{54.59}In_{44.66}Er_{0.75})_2S_{300}$ single crystals. Journal of Materials Science: Materials in Electronics. 2016;**27**:3258-3264. DOI: 10.1007/s10854-015-4153-2

[23] Kityk I, Yukhymchuk V, Fedorchuk A, Halyan V, Ivashchenko I, Oleksieyuk I, et al. Laser stimulated piezo-optics of γ-irradiated $(Ga_{55}In_{45})_2S_{300}$ and $(Ga_{54.59}In_{44.66}Er_{0.75})_2S_{300}$ single crystals. Journal of Alloys and Compounds. 2017;**722**:265-271. DOI: 10.1016/j.jallcom.2017.06.072

[24] Halyan V, Khyzhun O, Ivashchenko I, Kevshyn A, Olekseyuk I, Tyshchenko P, et al. Electronic structure and optical properties of $(Ga_{70}La_{30})_2S_{300}$ and $(Ga_{69.75}La_{29.75}Er_{0.5})_2S_{300}$ single crystals, novel light-converting materials. Physica B: Condensed Matter. 2018;**544**: 10-16. DOI: 10.1016/j.physb.2018.05.023

[25] Ivashchenkoa I, Halyan V, Kevshyn A, Kubatska T, Rosolovska V, Tishchenko P, et al. Physical properties of the $(Ga_{70}La_{30})_2S_{300}$, $(Ga_{69.75}La_{29.75}Er_{0.5})_2S_{300}$ single crystals. Acta Physica Polonica A. 2018;**133**: 994-996. DOI: 10.12693/APhysPolA.133.994

[26] Halyan V, Kityk I, Kevshyn A, Ivashchenko I, Lakshminarayana G, Shevchuk M, et al. Effect of temperature on the structure and luminescence properties of $Ag_{0.05}Ga_{0.05}Ge_{0.95}S_2$-$Er_2S_3$ glasses. Journal of Luminescence. 2017; **181**:315-320. DOI: 10.1016/j.jlumin.2016.09.022

[27] Kuznetsov A, Velazquez J, Tikhomirov V, Mendez-Ramos J,

Moshchalkov V. Quantum yield of luminescence of Ag nanoclusters dispersed within transparent bulk glass vs. glass composition and temperature. Applied Physics Letters. 2012;**101**: 251106. DOI: 10.1063/1.4772957

[28] Agafonova D, Kolobkova E, Sidorov A. Temperature dependence of the luminescence intensity in optical fibers of oxyfluoride glass with CdS and CdS_xSe_{1-x} quantum dots. Technical Physics Letters. 2013;**39**:629-631. DOI: 10.1134/S1063785013070158

Chapter 4

Optically Clear Adhesives for OLED

Joel T. Abrahamson, Hollis Z. Beagi, Fay Salmon and Christopher J. Campbell

Abstract

Optically clear adhesives (OCA) have been used for more than a decade to bond rigid LCD and AMOLED displays for consumer electronics applications, offering optical, mechanical, and electrical performance benefits. The performance requirements of an OCA to bond cover window, touch sensors, and circular polarizers in a plastic OLED display to bent cover glass or a flexible, foldable OLED display are drastically different from a flat, rigid device. For plastic OLED to bent cover glass bonding, the adhesive needs to be strong enough to resist spring back of the flat, plastic OLED devices. For flexible, foldable OLED displays, the neutral plane needs to be managed during folding keeping strain to a minimum in critical layers of the device (e.g., touch sensor, TFT, TFE), and the OCA cannot deform (or cause other layers to deform) during the folding process. Folding also brings challenges to touch sensors that can no longer use conventional passivation layers. As a result, the OCA will be responsible for preventing corrosion of touch sensor materials such as metal mesh, silver nanowire, carbon nanotube, and graphene. The chapter will discuss OCA performance requirements for rigid, flexible, and foldable OLED bonding.

Keywords: adhesives, OLED, optically clear, OCA, LOCA, displays, bonding, consumer electronics

1. Introduction

Optically clear adhesives (OCA) and liquid optically clear adhesives (LOCA or OCR for optically clear resins) have been used in consumer electronics, industrial and automotive bonding solutions for more than 15 years. OCA usage increased significantly as consumer electronics devices saw the transition from resistive to capacitive touch with its first implementation in the LG Prada phone in 2006 [1]. The advantage of using an OCA shows improvements in mechanical, optical and electrical performance of the display module and device. The initial application in OCAs in rigid OLED-based devices was similar to LCD devices, but as plastic OLED (pOLED) devices were introduced to the market, OCAs enabled new form factors, such as curved OLEDs in the Galaxy Round and Galaxy Gear S [2] as well as the Galaxy Note Edge [3].

There are two important mechanical considerations for optically clear adhesives – does it stick (adhesive strength) and how strong is it (cohesive strength). Adhesive strength can be defined by the work of adhesion, which is the amount of work required to separate the adhesion from the adherend. Adhesion can be achieved through van der Waals interactions, hydrogen bonding, acid–base, ionic, covalent and mechanical interlocking [4–6]. Film based OCAs are pressure sensitive adhesives, while LOCAs are liquids that are dispensed and cured. Pressure sensitive adhesives exhibit tack when the modulus is below 3×10^5 Pa or the so-called Dahlquist criterion [7] at the application temperature.

Once the adhesive sticks, the cohesive strength of the material determines the mechanical properties of the bond. Conventional adhesives are functional to the point of bond failure. For optically clear adhesives, they are functional to the point of cavitation that can occur at a significantly lower strength than bond failure [8]. The cohesive strength of an OCA can be designed to resist cavitation and bubble formation when it is warranted.

OCAs are often tested for reliability and durability at elevated temperatures and humidity (e.g. 65°C/90% RH). It is important to have sufficient adhesive strength to maintain performance through these conditions. Potential failure modes include moisture ingress at the interface and bulk plasticization of the adhesive. The adhesive should be selected to have sufficient strength based on the design of the module/device and the necessary durability for the adhesive.

Optically clear adhesives are typically based on acrylate chemistry, having a refractive index of 1.47–1.48 that is comparable to glass and other important materials (e.g. PET, polarizer, etc.) in a display device. By matching the index and using an adhesive to replace an air gap, this significantly reduces the amount of light reflected at an interface (~4% [9]). By using an OCA, both contrast and brightness can be improved in a device.

Reliability and durability of OCAs in elevated temperatures and humidity can have an impact on optical properties. The adhesive can absorb moisture under these conditions, becoming supersaturated at elevated temperatures. When the adhesive returns to ambient conditions, the excess moisture phase separates in the material becoming apparent as haze in the bulk of the material. This can be avoided by designing the adhesive to be anti-whitening [10]. Durability under UV exposure is a requirement in automotive and high performing consumer electronics applications. A common failure mode under these conditions is yellowing. Careful materials selection and formulation by the materials supplier can prevent yellowing by UV. Circular polarizers in OLED devices can reduce UV transmission to the OLED panel [11]. As POL-less devices [12] become a reality, the OCA will need to take on UV blocking functionality to protect the OLED device.

The electrical functionality of OCAs includes protecting sensitive touch sensors as well as enabling superior touch functionality. Industrial pressure sensitive adhesives often contain acrylic acid [13] to achieve good adhesion. Acidic species corrode touch sensors, such as ITO. Therefore, adhesives selected for bonding in optical applications must be acid-free. The potential module stack configurations for touch will be described in the next section. OCAs can be designed with an optimal dielectric constant at the touch sensing frequency (100 kHz) [14] from low ($D_k < 3$) to high ($D_k > 9$) to maximize the signal-to-noise ratio and minimize current and power to drive the touch circuit.

2. Key requirements and challenges in OLED optical bonding

The first OLED displays bonded in consumer electronics were rigid (**Figure 1a**) and sealed in glass barriers. While they cannot provide the promise of flexible form factors, the OLED patterning process can still make non-rectangular form factors such as rounded corners or notches, although LCDs can now also create these shapes of viewing areas. Rigid OLEDs also can provide higher contrast and wider color gamut compared to liquid crystal displays (LCDs), but the requirements for the adhesive are not dissimilar to those for rigid LCD optical bonding.

When OLEDs are manufactured on thin, plastic substrates like polyimide, they can enable new form factors that are bent or formed into curved shapes (**Figure 1b–d**). Now many different manufacturers are introducing dynamically flexible devices that can fold repeatedly, such as the Royole FlexPai [15], Samsung Galaxy Fold [16], and Huawei Mate X [17]. Besides these single-axis folding devices, the LG Signature OLED TV R9, announced in 2019 [18], demonstrates that even very large OLED modules can be constructed to roll open and closed from a cylindrical shape less than 30 cm in diameter. Although TVs are not often touch-enabled, this type of flexible device should still use optical bonding to add protective layers to the OLED.

Whatever the form factor, OLED display modules all require additional layers for durability, user interface, and optical functionality. At least one layer each of cover window (CW), touch sensor (TS), and circular polarizer (CP) is generally included, and these must be laminated together with OCA or LOCA for best contrast, brightness, and mechanical durability. The cover window is generally made of glass, although in a few cases manufacturers have used plastics like polycarbonate (PC), polyethylene terephthalate (PET), or clear polyimide (CPI). Plastic cover windows will not shatter and can more easily flex to meet the needs of dynamically foldable or flexible displays. Commonly an ink border or bezel, 10–60 μm thick, is printed around the perimeter of the CW to hide non-transparent supply circuitry in the TS and display layers.

The touch sensor itself may include a thin passivation layer over the functional layer (usually made of ITO, but sometimes metal mesh or silver nanowires) to protect the sensor from environmental aggressors. The CP may be positioned above the touch sensor (**Figure 1b**) to decrease the visibility of circuit patterns in the touch sensor. While **Figure 1** shows at most one separate TS layer, closely related to

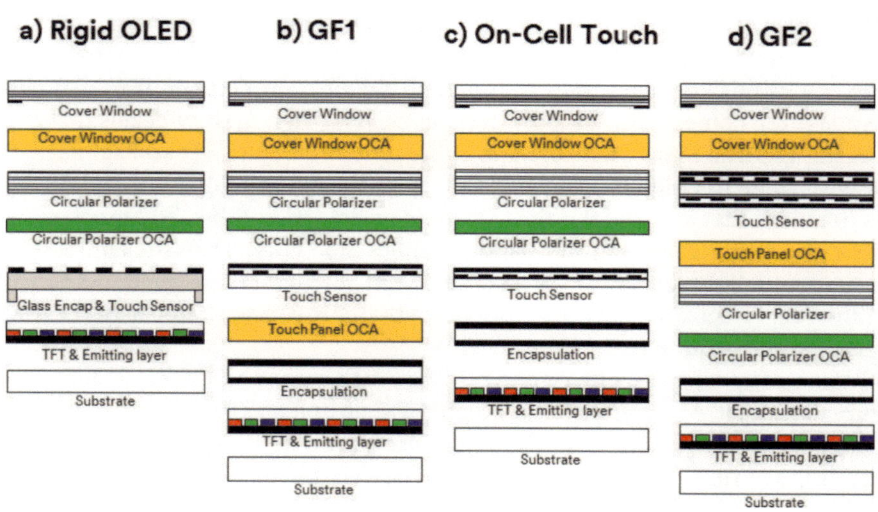

Figure 1.
OLED displays may be categorized according to the type of encapsulation and touch sensor used. (TFT = thin film transistor) (a) Rigid OLED, so named because the OLED is protected by an inflexible glass barrier with frits sealing the edges. Such designs may use a transparent conductor deposited on the glass as a touch sensor or use a separate film layer, as in (b) and (d). Architectures b-d use thin, flexible encapsulation (TFE), enabling these OLED to flex and bend. (b) GF1 means one glass cover and one film touch sensor layer. (c) On-cell touch, also called touch on encapsulation (TOE), is similar to (a) but with the sensor deposited on the flexible encapsulation layer. Therefore, the sensor substrate can be thinner than in GF1. (d) Another popular design emphasizing touch performance is GF2, which uses a single film layer with touch sensing layers patterned on both sides. (Graphic by Erik Iverson, 3M Display Materials & Systems Division).

GF2 (**Figure 1d**) is GFF (not pictured), in which two films comprise the touch sensor, each with one sensing layer, adhered with a thin OCA between them. Without separate sensor layers to resolve horizontal and vertical position, the touch sensitivity and precision of a GF1 design may be less than a GF2 design.

2.1 Optically clear adhesive requirements

As of 2018, OLEDs were used almost exclusively in smartphones and smartwatches, so special requirements for bonding larger displays (*e.g.*, notebooks, monitors) are less defined at present. The requirements and benefits of optical adhesives are most defined for the Cover Window OCA (often called OCA1) and Touch Panel OCA (OCA2 or OCA3, depending on the position of the polarizer). Either of these adhesives may also be a liquid resin (OCR).

Besides the general requirements for optical adhesives described in Section 1, OCA1 requires a balance between adhesive strength and capability to cover the step height of the ink bezel. Certainly strength, usually in terms of a high shear modulus, is critical to help the whole display resist impacts, a major source of user frustration every day when screens break. OCA1 is closest to an impact event on the cover window surface, but all adhesives bonded together in the display system will contribute to its resilience to impacts when the device is dropped.

On the other hand, the OCA must move sufficiently under the shear forces of lamination but then hold its position and shape afterwards, even through temperature cycles of regular use. As bezels are often now less than 1 mm wide, the OCA must bond over an ink step that may be less than 10 times longer than it is high – a challenging aspect ratio to cover without trapping an air bubble at the base of the ink step. Furthermore, the OCA must be strong enough to resist temperature cycling and moisture exposure that could pull it away from the ink step, creating edge bubbles. Bending the display around a curved edge only increases the stress on the adhesive and magnifies the challenges.

The inherent viscoelasticity of OCAs as pressure-sensitive adhesives enables good step coverage of ink or flexible printed circuit (FPC) bond lines. However, as design trends push displays to be thinner and thinner, the solution for great step coverage often comes with UV post-curable OCAs, when narrow bezels make LOCA impractical. Such OCAs are applied in a more viscous, yet still solid, film form with some initial tack. They can flow with the pressure of lamination to cover steps ink steps that are 30–50% the thickness of the OCA. A heated autoclave process step may further improve step coverage capability. Then the laminate is irradiated with light, generally in the UVA spectrum, to increase the bonding within the adhesive, effectively "locking" it in place to improve its durability to the level of a high-modulus adhesive.

In most designs, especially those with the touch sensor on the encapsulation (**Figure 1a, c**), OCA1 benefits from a higher D_k, as much as 9, to increase capacitive coupling between the finger and the sensor. However, designs with a separate touch sensor (**Figure 1b, d**) often require a *lower* D_k, < 3, for the touch panel OCA to decrease coupling of electromagnetic interference (EMI) noise from the display. In either case, however, stability of D_k is important, over operating temperatures and humidity (moisture absorption generally increases Dk) and through the life of the device, as an accelerated aging test would simulate. $\Delta D_k < 0.3$ is a good guideline, and higher performance applications may require a narrower range.

These properties benefit reliability, durability, and esthetics – all values to the end user of the device. On the other hand, manufacturers and the display supply chain have other requirements. To improve overall process yield and efficiency, these players desire re-workability: targeted or triggered failure of the adhesive cleanly and without residue from one interface. Then components may be separated; expensive parts, especially the OLED and the cover glass, may be recovered. The challenge is that the re-work condition – be it heat, cold, light exposure, or some other factor – must be outside the realm of normal operating conditions so that the display does not de-bond prematurely after it leaves the factory line.

Dimensional stability is another property more visible to the manufacturers than to the end user. OCAs may be delivered as master rolls, and LOCAs will be dispensed or printed as a liquid, but for aggressive tolerances and more complex shapes like rounded corners, manufacturers often opt for individual pre-cut adhesive parts. In that case, the cut OCA must maintain its shape and dimension through environmental conditions of shipping and duration of expected shelf life. The flow, creep, or oozing of the adhesive at elevated temperatures must be limited; consider that the inside of an uncontrolled shipping container can easily exceed 50° C. Similarly, the shipment and packaging format interplays with the formulation; the sustained pressure of a stack of cut parts should also cause only minimal oozing. These properties for cut-part stability are often at odds to the viscosity needed for an uncured PSA to cover steps, as described earlier.

Most OCAs for display bonding are designed to resist whitening, as described in the previous section. Some degree of water resistance has been valuable in OCAs, but the next frontier may be resistance to a broader range of chemical ingress, especially for designs in which the bezels are very narrow or non-existent. Wearable displays also encounter more splashes compared to phones, which may often be partly protected inside a pocket or bag. Normally, a rim tape or lens bonding adhesive may adhere the cover window to the device housing underneath the bezel, and materials chemists may use a wider range of polymers to block chemical ingress, since the material does not need to be optically transparent. The challenge is heightened since everyday chemicals that might contact screens include oils/fats (cooking oil, sweat), alcohols, acids (soda, vinegar), and bases (ammonia, cleaning products) – both hydrophobic and hydrophilic substances.

While manufacturers generally select OCAs for OLED separately from the layers they bond, the CP OCA is widely sold with the CP itself, a different supply chain. Thus, low cost for the integrated product is key, in addition to technical requirements of good adhesion to the CP at a thickness of $\leq$25 μm. Since the polarizer absorbs a large fraction of UV light, this OCA cannot practically be UV post-curable. As future CPs continue to decrease in thickness [19–21], with thinner protective layers around the optically active layers, the CP OCA may need to serve a greater role in protecting the polarizer from harmful environmental factors.

2.2 Bonding requirements in bent or curved displays

The performance requirements of an OCA to bond layers in a flexible OLED display are quite different from a rigid, flat device. The neutral plane must be balanced to minimize strain in critical layers of the device. Thin film transistors (TFTs), thin flexible encapsulation (TFE), and touch sensors are predicted to be most vulnerable, although newer silver nanowire and metal mesh sensors are more flexible than ITO [22, 23]. Although manufacturers continue to improve the durability of OLED displays, publicly available literature indicates that critical components are still quite sensitive to strain. The SiO_2 layer of the TFT on polyimide may break at 0.2% tensile strain [24] and the TFE at around 0.3% in bending [25].

When the form of the device requires the OLED to bend only once (during manufacturing), the cover window can be made of glass, which offers the best resistance to scratches and protects the display from impacts to the screen. However, the modulus (*i.e.*, stiffness) of the glass is much higher than the other plastic films of the display (polarizer, touch sensor, OLED), whereas the plastic films feel a "spring-back" force due to their tendency to relax back to their original flat form. Since the glass will not move or deform, the OCAs in the curved display module must have very high adhesion to hold the plastic films in their curved state; otherwise they will pull away from the glass at the bend and leave bubbles.

Examples of OCAs that have the strength needed for durable curved bonding are 3M™ Contrast Enhancement Films (CEF) 30xx and 31xx, where xx denotes the thickness in mils. **Table 1** describes one important property, peel adhesion, for these adhesives. Their primary difference is that CEF30xx is UV post-curable for step coverage, as described above, and its samples in the table were cured with 1 J/cm^2 UVA prior to peel testing.

2.3 Bonding requirements in dynamically flexible displays

To minimize strain on a flexible OLED display, an adhesive should either perfectly mechanically decouple the layers from each other in bending (*i.e.*, with very low modulus) or adjust the position(s) of the neutral plane(s) during the bend. Plastic OLED displays have some inherent flexibility and have been demonstrated to fold thousands of times to a radius of 4 mm or less [26, 27], but the right adhesives are required to integrate the display layer into a durable, functional device while maintaining that flexibility. Besides balancing the neutral plane(s) to minimize strain on the OLED, when the device is folded closed, the adhesive should have minimal creep to avoid flowing and changing in thickness in response to stress. Upon unfolding the device, no buckles or wrinkles should be visible, and the adhesive should assist it in recovering quickly to a flat, smooth profile. Finally, all these adhesive properties should vary as little as possible over the operating temperatures and conditions of the device. These requirements point to a soft, elastic material with a low glass transition temperature (T_g), below the minimum operating temperature. If operating temperatures often extend down to $-20°C, T_g < -30°$ C would be a minimal requirement.

The elasticity requirement contrasts with the common viscoelasticity of OCAs in static applications, and a soft (low modulus) adhesive would typically also sacrifice adhesion. Practically speaking, the foldable OCA should still maintain peel adhesion of 1 N/cm at a minimum. The surface energy and cleanliness of the interfaces with the adhesive also affect adhesion, however, so this property may be improved with plasma treatment or primer layers to increase surface energy.

When a flexible display is expected to bend thousands of times, a foldable OCA should retain these mechanical properties with minimal shift or hysteresis through repeated stress/strain cycling. Besides testing in application, the adhesive properties may be characterized through tensile, compressive, and shear loading, for example,

OCA Name	*Thickness (μm)*	*20 min dwell*	*3 d dwell*
CEF3004	100	11.2	14.0
CEF3006	150	15.1	17.4
CEF3104	100	11.0	14.1
CEF3106	150	12.0	15.2

ASTM D3330 modified test method, 1-cm-wide strips, 305 mm/min peel rate, 2.0 mil polyester backing. All dwells at 23°C/50% relative humidity.

Table 1.
180° Peel adhesion (N/cm) from float glass for curved OLED display OCAs after varying dwell times.

using mechanical load frames, rheometers, or texture analyzers [28]. Some mechanical testing to support the simulation of flexible adhesives is described later in Section 4.

Even a folding radius of 5 mm can require shear strain of >300% in the adhesives during folding (depending on the thickness of the various display layers). Foldable adhesives must certainly shear to this extent without breaking or permanently deforming but are also expected to release strain quickly when stress is removed upon unfolding. However, recovery time of the flexible display also greatly depends on the properties of the other film layers, which have much higher moduli, so it is difficult to distill the recovery requirement for the system into a generic recommendation for the adhesives.

2.4 Future requirements for optically clear adhesives for OLEDs

Many trends in OLED design focus on enabling greater flexibility, and new adhesives will be needed to support this as well. Through the next decade, manufacturers will push for narrower folding/rolling radii, multiple folding axes in displays, and longer device lifetimes, especially in terms of cycles of flexing. As folding radius decreases below 3 mm, the required shear strain will likely increase from 300% to roughly 500 to 700% or more. Additional folding axes will also increase the total magnitude of shear displacement, as each axis adds its own displacement to the system. A rollable display can be considered as the limit of many axes of folding across the length of the device. Rolling devices will initially launch with radii of several centimeters, which is larger than today's one-axis folding displays, but the shear will add with each bend such that the "creep" with applied stress will be a more critical property. With larger creep, even a flexible adhesive could shear out several mm at the edge when the display is rolled, resulting in unsightly uneven edges or, worse, display layers pressing into the bezel/frame bonding and generating further stress.

Thinner display modules will assist flexibility, so trends in integrating functional components are likely to continue. These may include coated circular polarizers [19–21] or color filters integrated with the OLED encapsulation [12]. With thin flexible cover windows, expect adhesives on either side of the OLED display to play a greater role in protecting the module from impact damage. This will be especially true for plastic CWs compared to glass, since they have lower elastic moduli.

Plastic cover windows may see further growth in automotive displays, but for different reasons. Some manufacturers and safety authorities have been concerned about the possibility of display cover glass shattering in an impact or crash and thus have preferred plastic CW for safety reasons. High-performance plastic CWs are often made of PC or PMMA (poly[methylmethacrylate]), so durable bonding to these low surface energy plastics, with their tendency to release gases under heating, will be important to future automotive OCAs. The reliability and accelerated aging tests for transportation applications are generally more severe than those for consumer electronics. The displays must withstand extremes of heat (often −40 to +110°C), humidity, and light (especially UV) irradiation, but also conditions like sand for desert environments and salt spray on the coasts. Finally, the interiors of many concept cars already feature large displays on the complex curves that designers favor for esthetics and ergonomics. Expect curved displays to grow in size and in popularity in production models in the coming years.

Although it takes time to decrease costs and prove reliability, gradually OLEDs will expand into outdoor applications as well, as consumers begin to expect a certain level of display quality from their everyday interactions with personal and mobile devices. These outdoor bonding applications will demand reliability from the OCA, like the automotive market. Transparent OLED displays are still in their infancy and will likely require lower levels of optical absorptivity, haze, reflectivity, and color for all layers, including optical adhesives.

A different design trend in OLED displays that has already begun in today's devices is the integration or hiding of the many cameras and sensors that now support the devices' interactions with their users and environments. The goal is always to expand the active area to occupy an ever-greater fraction of the viewable front of the device. Additionally, a camera behind the display provides a more natural interaction in video calling; the camera may track a user's gaze at the same location where the user views the other person's face, giving the impression of eye contact. The implication for bonding with OCAs is stricter dimensional control and tolerance requirements, extending to non-rectangular shapes. Already many sensors and cameras can fit in a droplet or notch shape cut of out of the display and adhesives at one edge of the screen. When the camera(s) are behind the display, some adhesive layers must precisely maintain a small hole in the middle of the part, as an aperture for the camera.

3. Folding test method development and differentiation of multi-layer concepts

3M has been investigating fatigue and shear performance of polymer film substrates and multi-layer stacks consisting of OCAs (Optically Clear Adhesives) and polymer films in response to dynamic bending. This work was prompted, in-part, to support materials and systems development in the realm of foldable OLED (Organic Light Emitting Diode) market technologies and applications. One device that is used to study the dynamic bending performance of these films and film stacks is an internally-designed and built bend tester. Samples can be mounted to the bend tester and folded to a specific bend radius and number of bend cycles at a defined cycle rate (bend cycles per minute). From a materials perspective, failure in the form or yielding, breakage and buckling can occur during testing. In actual flexible device constructions, the light emitting and barrier layers can be compromised over certain stress and strain thresholds. These failures are influenced largely by the stress or strain amplitude as the samples are placed in the folded conformation of interest. In this section, we will discuss the methodology behind development of a controlled and repeatable folding test that minimizes bending stress on foldable test specimens, types of defects observed in single and multi-layer polymeric stacks consisting of OCA, and additional methods of concept differentiation by incorporating ITO (Indium Tin Oxide) coated PET (polyethylene terephthalate) into the multi-layer test specimens.

3.1 Investigation and definition of test method parameters

3.1.1 Hinge or pivot design

One of the first design parameters considered during initial folding test method development is the hinge design (or pivot locations) of the test apparatus. It is important to have at least some understanding of the impact of hinge design selection. Some types of hinge design can impart added strain (or stretching) on samples during the process of folding. With single hinge folding design, or folding with only one pivot point, there will be one or more critical angles in which the test specimen is stretched beyond its original length. This type of tester design is undesirable because the design of the tester can damage the samples, and will introduce added variables, such as test specimen attachment location and attachment adhesive properties, that will dictate how much overall strain is imparted to the sample during the action of folding and unfolding. Slight changes in attachment location and attachment adhesive properties will have an impact on variability and reproducibility of the test results. Note, this logic can be applied to certain mandrel bend folding designs as well, if the strain on test specimens is not actuated to prevent sample lengthening or stretching during folding. For this reason, we have defined a dual pivot testing apparatus (**Figure 2**), which can prevent added strain or stretching of the samples during folding.

3.1.2 Attachment method and location

In many cases, test specimens will consist of alternating layers of polymer film and adhesive. The primary roles of the adhesive in such constructions will be to adhere the layers together and to prevent added strain on sensitive layers, such as the display or fragile coatings. The mechanical benefit of the adhesive is maximized when specimen layers can freely move and shear across one another during folding and unfolding, thus minimizing stress on fragile layers. For this reason, attaching specimens to the test apparatus using adhesive (instead of clamping) is recommended, especially if it is representative of actual folding device construction. If specimens are attached to the folding test apparatus with clamps, there will be new test variables to consider, such as clamping pressure and location.

When attaching specimens to the folding apparatus using adhesive or tape, the next decision becomes where to attach the test specimen. Samples attached too close to the apex of the fold can add strain on the samples, due to lateral pulling of the samples to the mounting plates during folding. This can also add undesired variability to the test response, in addition to added strain on test specimens. Test specimen shape in the folded configuration can be measured or modeled. For our folding test method, we have defined an attachment location (denoted as L_{free} in **Figure 3b**, below) not closer than 3*g/2, where g is the distance between confining test plates and attachment adhesive.

3.1.3 Test specimen misalignment

Misalignment of the test specimen can be caused by off-centered attachment of the test specimen, misalignment of the pivot points (non-parallel or collinear; **Figure 4**), or curved/warped mounting plates. Such forms of misalignment can result in different patterns and variability in the test response (from left to right sides of the test apparatus, for example; **Figure 5**).

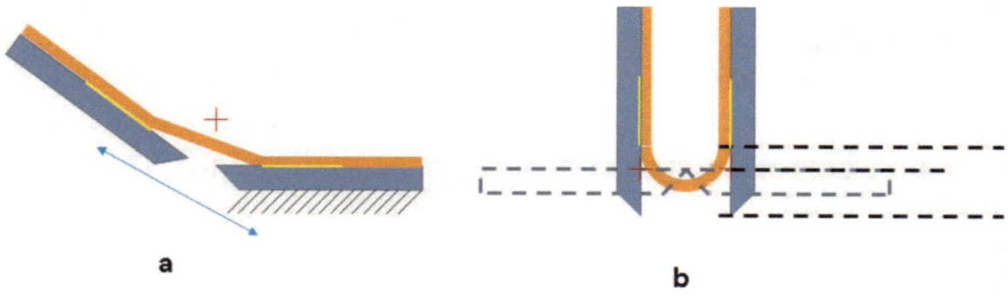

Figure 2.
(a, b) Specimen geometry during folding for a single vs. dual pivot test apparatus. Tester mounting plates are depicted in blue, attachment adhesive is depicted in yellow, test specimen is depicted in orange, pivot locations are depicted by '+' signs. (a) shows a single pivot hinge design, in which samples can be stretched at a critical closing angle during testing. (b) shows a dual pivot hinge design, which can eliminate added stretching of the test specimen during folding. (Figures by Tom Corrigan, Ph.D. 3M Corporate Research Systems Laboratory).

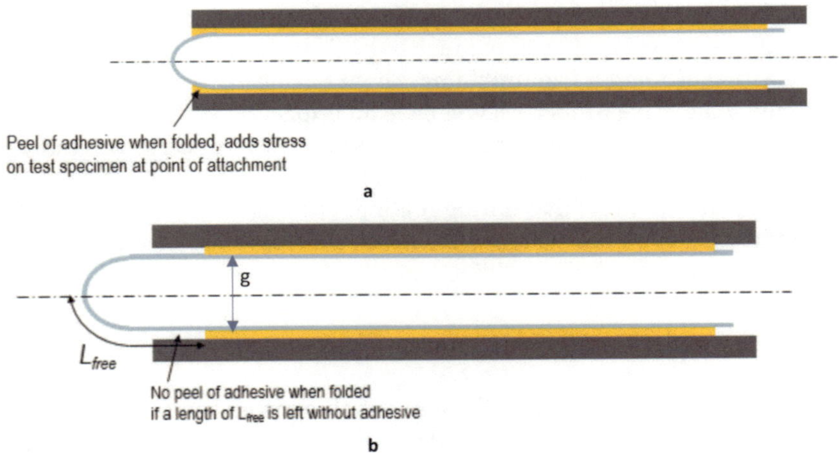

Figure 3.
(a, b) Specimen geometry denoting location of attachment location with bond all the way to edge of substrate backing in (a), while (b) depicts a region without adhesive contact near the fold region.

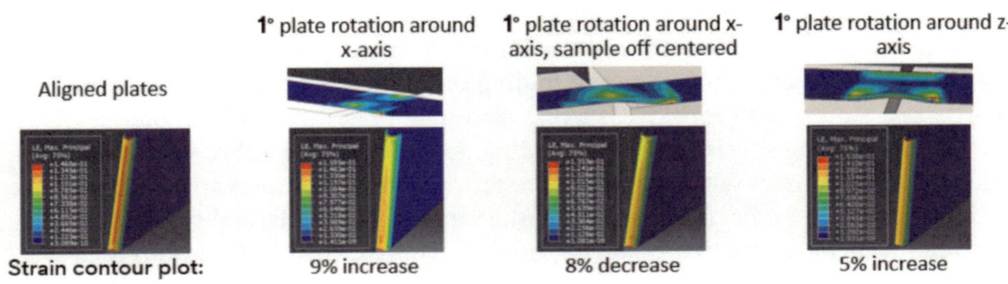

Figure 4.
Shown above are modeling results, which depict changes in strain on test specimens depending on their alignment on the test apparatus. (Courtesy of Samad Javid, 3M Corporate Research Process Lab).

3.1.4 Sample conditioning and preparation

Both temperature and humidity can affect the test response, particularly for multi-layer test specimens that include adhesive in some layers, since some adhesives may be quite sensitive to moisture uptake. Dependent on the substrates used, this water uptake could apply to films, as well. Specimens should be conditioned at well-defined environmental parameters prior to testing. Testing should also take place in a controlled temperature and humidity environment. Doing so can help improve the repeatability and reliability of the test.

Lamination conditions, sample edge quality and surface treatments that impact adhesion can also influence the test response. Differing amounts of stored stress introduced during manufacturing, and tension applied during the lamination and sample preparation process can change the starting stress states of a test specimen, which can then influence the stress profiles during folding and unfolding.

3.1.5 Defects observed

Depicted below are several types of failure modes observed during bend testing of single and multi-layer test specimens. Crazing (**Figure 6a**) is caused by slippage of crystal structures and void formations in polymeric films. Local buckling (**Figure 6c**) is caused by adhesion failure between layers. Global buckling (**Figure 6d**) is caused by severe instability, which can result in inverted folding of the test specimen.

DOI: http://dx.doi.org/10.5772/intechopen.88659

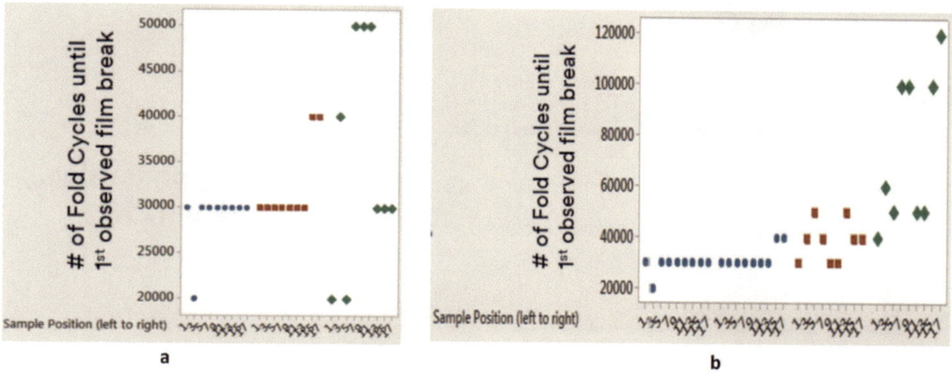

Figure 5.
(a, b) Results from testing of 4mil low melt PEN (polyethylene naphthalate) at a nominal gap size of 4 mm. Sample position is measured in inches from left to right side of test apparatus. Samples were checked every 10,000 fold cycles. (a) shows the difference in test response observed when test plates are very flat (2 test cycle replicates are depicted on this plot by red and blue), vs. curved with variation of folding gap by only 0.05 mm as measured with a feeler gauge (data depicted as a general smile pattern in green). (b) shows a sinusoidal test response (plotted in red and green) as a result of loose bearings design, which control the motion profile of the test plates. Tighter bearings design, in this case, yield a flatter more controlled test response profile (plotted in blue).

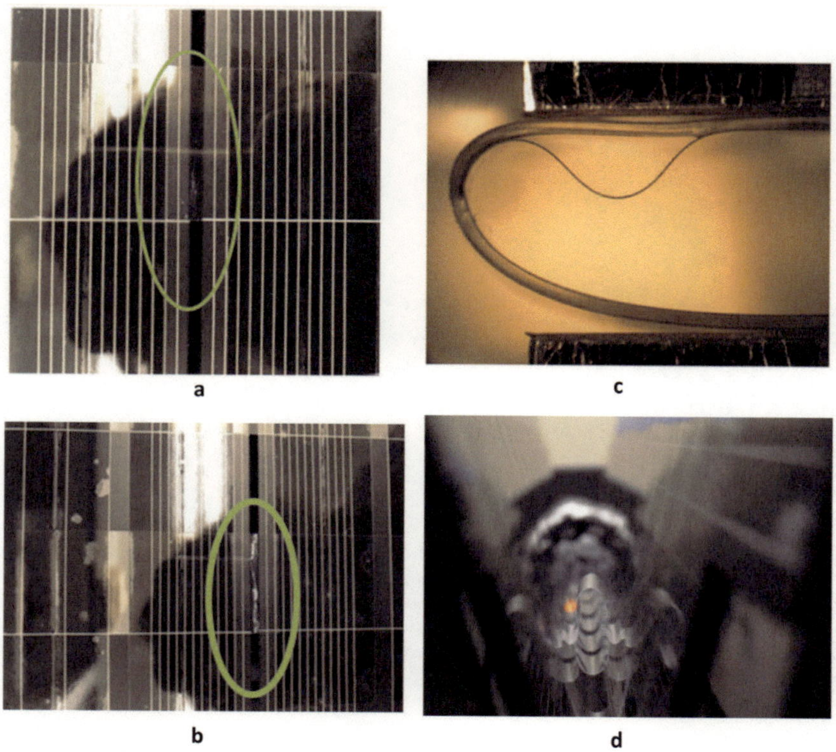

Figure 6.
Depiction of typical failure modes in foldable testing, including substrate crazing (whitening) (a), breakage (b), local buckling (c) and global buckling (d).

3.2 A test for OCA decoupling and neutral plane management

There is widely available literature to explain the effects of decoupling and isolating multiple neutral planes in a multi-layer construction [14, 29]. In the ideal

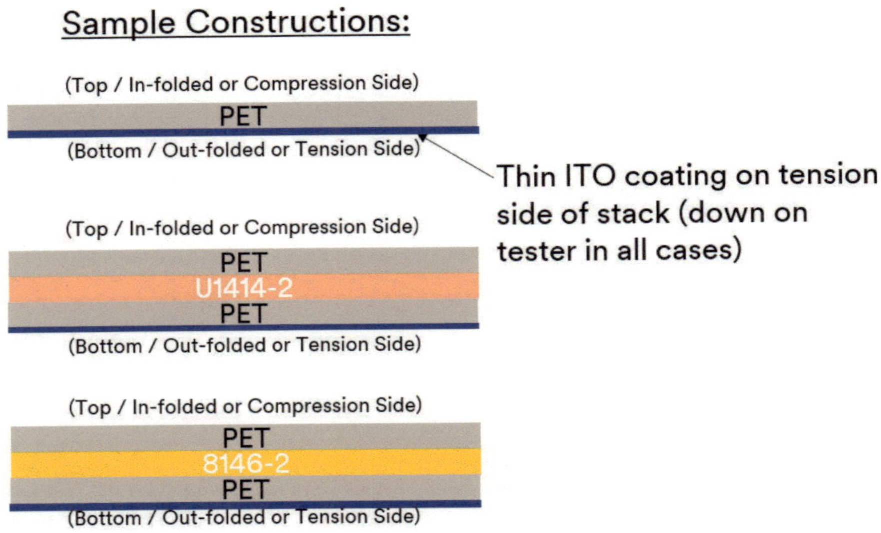

a

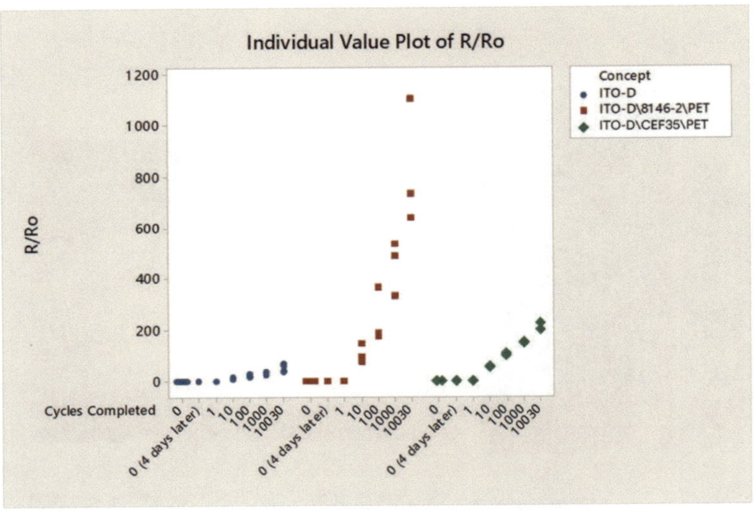

b

Figure 7.
(a) shows descriptions of test specimens comprising a single layer of 2mil PET, and 3-layer samples comprising of 2mil 3MÔ OCA 8146–2 (a standard 3M OCA) and 2mil 3MÔ CEF35 (a 3M foldable OCA). (b) Results show that less resistance increase is observed for the test specimens comprising of 3MÔ foldable OCA, CEF35, than for specimens containing standard 3MÔ OCA 8146-2. These results suggest that the strain amplitude on the tensioned ITO coated side of the test specimen is reduced when using 3MÔ foldable OCA, thus reducing damage to the fragile coating.

case, the neutral planes of each layer in the composite construction are preserved by frictionless slip (or perfect decoupling of each layer). In reality, the lay-up construction (modulus and thickness of each layer), in addition to assembly (such as tension during the lamination process) of the test specimen or display stack will influence stress profiles. 3M Foldable OCA has a lower modulus over a wide range of operating temperatures, as compared to standard commercial OCAs. This offers greater ability to decouple the stiff layers in a flexible display module and reduce the overall bending strain [29, 30].

A simple experiment was performed to assert the performance of 3M™ OCA CEF35, a 3M Foldable OCA. A thin layer of ITO was coated onto 2mil PET. The ITO coated PET was used to create 3-layer test specimens, comprising either a standard OCA 8146–2 or CEF35 at the same thickness of 2 mils (**Figure 7a**). The test specimens were folded, with ITO on the out-folded/tension side, using a dual pivot bend tester. The folding rate for this test was 30 cycles per minute, with nominal folded gap size of 4 mm. Samples were conditioned for 24 h and tested at 23C, 50%RH. Electrical resistance of the ITO coating was measured at the end locations of each test specimen (using an Ohm meter) after 0, 1, 10, 100, 1000 and 10,000 fold cycles. Results (**Figure 7b**) showed significantly less resistance increase for stacks constructed with CEF35 than for those containing OCA 8146–2, proving experimentally that CEF35 more effectively decouples the polymer film layers, due to a lower shear modulus (G') and higher yield under minimal load, thus, causing less damage to the conductive layer.

3.3 Folding test method summary

In conclusion, there are several factors to consider when designing a reliable and repeatable folding test method. Factors such as hinge or pivot design, attachment location, sample and fold axis alignment, as well as sample preparation and conditioning, and surface treatments can all have an impact on the reliability and repeatability of the test. Once these factors are controlled, the folding test can be an effective tool for use in differentiating multi-layered constructions representative of foldable displays. IEC standard 62715-6-1 [31] can be referenced and calls out several different methods of bending deformation that can be used to evaluate Flexible Display Devices. Common failure modes include crazing, breakage, local and global buckling. Most, if not all, of these failure modes can be influenced by the properties of the adhesive used to construct the test specimen. 3M Foldable OCA can help to decouple the polymeric film layers of a multi-layer test specimen, and therefor reduce strain on fragile coatings and components, due to its low modulus as compared to standard OCAs, such as 3M OCA 8146.

4. Modeling of foldable OLED panels

For foldable AMOLED (OLED) display panels, the repeated folding and unfolding create new mechanical requirements and challenges beyond those of fixed displays. In most applications, the multiple functional layers of a foldable display are bonded by optically clear adhesives (OCA) to form a multilayered thin panel or film stack that is flexible to allow folding of the panel.

The folding of a multilayered panel will introduce bending stress and shear stress not only within the individual layers but also between the bonded layers. Depending on the panel layup design, these bending stresses can reach very high levels causing various failures including fracture of individual layers or sensitive components within a layer, delamination, and buckling. Understanding the nature of these stresses will help in producing robust designs that ensure reliable folding performance.

Because the folding of a display by nature is a problem of large deflection of a multilayer film stack, the classical beam bending theory such as the Euler–Bernoulli approach will be limited in its ability to describe accurately the bending stress and interlaminar stress. The thin nature of the OLED display also makes it difficult to experimentally measure the stresses or visualize the deformation of the layers within the panel. Modeling then becomes an effective tool to provide insights into the mechanical folding behavior. Numerical analysis methods, such as finite element analysis (FEA), offer detailed information of stress and related deformation in each layer of the film stack, and allow virtual testing to evaluate different design scenarios [14, 32–35].

In this chapter, we will provide an introduction of the folding simulation of OLED display panels. The aim is to demonstrate the basic process of modeling the folding and the potential failure modes of foldable displays to help design engineers to become familiar with:

- The interactions between layers of the display film stack with its bonding layers, the OCA layer, and the role of OCA on the folding performance of the film stack.

- Some basic modeling methods to simulate the folding of the film stack and the OCA.

As general guidance, we will use modeling examples of a simplified but representative display film stack to describe and discuss the models and the results.

We will also focus on the roles of the soft bonding layer, the OCAs, in the discussions of the modeling results. We will examine the overall performance of the film stack to provide basic understanding of the effectiveness of a soft bonding layer in reducing bending stress and mitigating related potential failure modes. This understanding can be valuable because the material properties and thickness of the functional layers in a display, such as the OLED, circular polarizer film, and the touch sensor film are designed to achieve their specific optical or electronic functions, and the options to vary the thickness and material properties of these functional layers may be limited. Managing a film stack's neutral plane by employing thicker or stiffer supplementary layers may be undesirable if the display panel is to be flexible. On the other hand, when the display panels become thinner, this imposes its own set of challenges in durability and usability. Optimization of the display film stack through the design of the bonding layer becomes an attractive and often a necessary option.

4.1 OLED panel and construction

The typical functional layers of an OLED film stack may include: cover film, the touch sensor layer, circular polarizer, thin film encapsulation (TFE) layer, the AMOLED (OLED) display unit, the back sheet, substrate, and others. The OLED display unit itself also consists of multiple components including a flexible substrate, encapsulation/barrier, thin film transducer (TFT), and transparent electrode. In more recent developments, display manufacturers are also pursuing integration of some of these components for thin form factors with fewer layers. The OCAs are placed between layers to bond them into an integrated display film stack that is flexible to allow folding as shown in **Figure 1**. The common industry practice for specifying the extent of the folding is by the radius of curvature of the folded region.

4.2 Design requirements for foldable multilayered display panel

The folding of a multilayered panel will introduce bending stresses, both in-plane tensile or compressive stresses and shear stresses between the layers. While each individual layer is flexible to withstand being folded into a small radius of curvature, the bonded multilayer panel will exhibit a much higher stiffness due to the larger thickness of the whole panel, and potentially generate very high bending stresses. The high bending stresses may cause failure in various forms. Of the various components and layers in an OLED display panel, the permeation barrier layer, the indium tin oxide (ITO) layer, and the oxide dielectric layers of thin film transistors (TFT) are sensitive to tensile stress and strain, and they can fracture at a relatively low tensile strain [24, 36].

While there are many other design requirements for a display unit, in terms of folding mechanical performance, flexibility, integrity, and durability without damage are required, and fracture, delamination, or buckling under repeated bending must be prevented.

Other design requirements, including impact resistance to drop impact of display device or ball/foreign object impact, and scratch resistance are also critical for foldable displays. They are interrelated with the foldability performance, and often have competing design requirements. A more systematic and holistic design approach is necessary to incorporate all the aspects of design to achieve balanced optimal performance with some trade-offs, if necessary, among the various competing design requirements. Various modeling work has been performed in these areas at 3M, however, this belongs to a more expansive scope beyond that of the current chapter. For a more focused discussion, this modeling section will be limited to the topic of integrity for foldability.

In the following sections, the general practice of finite element analysis for simulating the folding of a film stack will be discussed first, followed by a modeling example to illustrate the potential failure modes of the film stack that need to be addressed in the design of foldable displays and the effects of OCA in mitigating these failure modes.

4.3 Finite element analysis of folding of display film stack

4.3.1 Finite element analysis models

The thin film nature of the film stack makes it suitable for using a 2D model to represent the stack without losing much fidelity, especially when edge effects are not significant. To model the interactions among the different functional layers, each layer should be modeled as an individual part of the film stack. In particular, this simulates the deformation behavior that is critical in understanding the effects of OCA on the folding performance of film stack. The finite element mesh size should be sufficiently fine to resolve the stress gradient of the critical layers both in their thickness direction and in the in-plane direction, especially in the bend region where stresses vary through the thickness direction.

The bonding of the OCA to film can be simulated by various interaction definitions such as nodal constraint or node coupling of the mesh of the OCA layer to the mesh of the substrate film layer. This approach will typically not allow the simulation of the separation or debonding of the OCA from the substrates, however, it can offer most of the needed information to evaluate the bond performance without the added complexity of simulating the debonding process. To simulate the debonding process, cohesive elements may be used to represent the OCA bond. If cohesive elements are used, care should be taken to incorporate the cohesive law that describes the debonding force-displacement of the OCA bond attached to the specific films of the actual application because the bond strength depends on the specific substrate films. Also, separate cohesive laws for tension (normal to bond interface) and shear should be used to accurately represent the OCA bond behavior which will require calibrations of the cohesive laws from test data such as those from butt joint or T-joint tests and lap shear tests. In general, the correct application of cohesive elements or cohesive contact for OCAs is more involved. Proper training on the technique will help to avoid misrepresentation of OCA bonds. It is recommended that users consult with 3M technical support for test methods or test data related to bond strength simulation.

4.3.2 Modeling OCAs

Being a type of pressure sensitive adhesive, OCA needs to be soft enough to flow under application of pressure to wet the substrate in close proximity. This intimate contact between OCA and the substrate allows molecular interactions such as van der Waals forces to form the basis of the adhesion that underwrites the overall bond strength. This flow-like behavior of OCA can be described by a combination of nonlinear elastic and viscoelastic/viscoplastic material models.

For OCA's viscoelastic behavior, its relaxation function can be defined in terms of a series of exponentials known as the Prony series. For the nonlinear elastic or hyperelastic constitutive models, the models such as the polynomial hyperelastic models are most commonly used. They include the neo-Hookean model, Mooney-Rivlin model, and Yeoh model. All are effective in describing OCAs' ability to be stretched to very large strain. Other forms of hyperelastic models including Arruda-Boyce and Ogden models may also be used. The commercial finite element analysis packages, such as ABAQUS, ANSYS, and COMSOL offer the finite strain computational methods and material model libraries that allow advanced simulation of OCAs. Users can apply the material model calibration tools provided by the finite element analysis software packages to evaluate the material parameters for these constitutive models from OCA material characterization test data. It is also recommended that they contact 3M technical support for information on OCA material data for their specific applications.

4.3.3 Material characterization tests

Several material characterization tests are needed to generate data for the calibration of the constitutive models mentioned above. For calibrating the elastic or hyperelastic material models, one or more of the tests listed below needs to be performed on the OCA:

- Uniaxial tensile
- Simple shear
- Biaxial tensile
- Compression and/or bulk compression

For viscoelastic model calibration, the dynamic mechanical analysis (DMA) test can be performed to produce master curves of the storage modulus and loss modulus of OCA to calibrate the Prony series. Otherwise, relaxation or creep test data at various strain rates and temperatures of interest may be used for the Prony series calibration.

For calibration of OCA bond strength models, such as the cohesive zone model, the proper measurements of the bond strength using T-joint specimens and lap shear specimens should be emphasized. These tests should be performed on the actual substrate materials of the film stack, especially if the potential failure mode may be adhesion failure. 3M Technical Support can be a source for information related to the test methods for adhesive characterization.

4.4 Examples of finite element analysis of folding performance of a foldable display film stack

The following is an example of finite element analysis of dynamic folding simulation of a display film stack, followed by discussions of the analysis results to assess the folding performance.

In the finite element analysis (FEA) presented here, a simplified 7-layer film stack was used to represent a foldable OLED display (**Table 2**). The film stack was assumed to be attached to two rigid back plates joined by hinges. The center portion of the film stack was not attached to the rigid back plate to allow for folding of the stack. The folding action simulated here is the so-called 'out-folding' such that the film stack is folded towards its back layer. In the fully folded configuration, the space between the straight sections of the outer layer is 10 mm resulting in the bend section of the film stack forming an approximate semi-circle of radius 5 mm as shown in **Figure 8**. In the simulation, the film stack was folded in 3 s, then held in the folded configuration for 24 h before it was unfolded in 3 s. The material properties and the thickness of each layer used in this work are listed in **Table 2**. The total thickness of the film stack simulated here was 0.475 mm (layup 1) and 0.275 mm (layup 2). The simulation was repeated for the two film stack layups bonded by 3M™ Foldable OCA CEF35 and 3M™ OCA 8211. 3M™ CEF35 has a lower storage modulus than that of 3M™ OCA 8211, and exhibits a more elastic behavior as indicated by its lower Tan delta value (**Table 2**) than that of 3M™ OCA 8211.

The simulation was performed using the ABAQUS finite element analysis package. Due to the large ratio of film stack panel width to thickness, a 2-dimentional

Layers	Thickness (μm)	Material	Modulus of elasticity, E, (GPa)	Storage Modulus, G', (GPa) at frequency of 1 rad/s	Tan Delta
Cover film	150 (50)	PET	3.5	NA*	NA
OCA	25	3M™ CEF35 (OCA 1) 3M™ OCA 8211 (OCA 2)	/	0.03E-3 0.07E-3	0.33 0.70
Circular polarizer	75 (125)	Triacetate	3.2	NA	NA
OCA	25	3M™ CEF35 (OCA 1) 3M OCA 8211 (OCA 2)	/	0.03E-3 0.07E-3	0.33 0.70
AMOLED	75	Polyimide substrate	3.5	NA	NA
OCA	25	3M™ CEF35 (OCA 1) 3M OCA 8211 (OCA 2)	/	0.03E-3 0.07E-3	0.33 0.70
Back plate	100 (150)	PET	2.7	NA	NA

**NA = Not Applicable*
The thickness shown in brackets in the table are the values used for stack layup 2.

Table 2.
The simplified 7-layer display film stack layup and the properties used in the simulation.

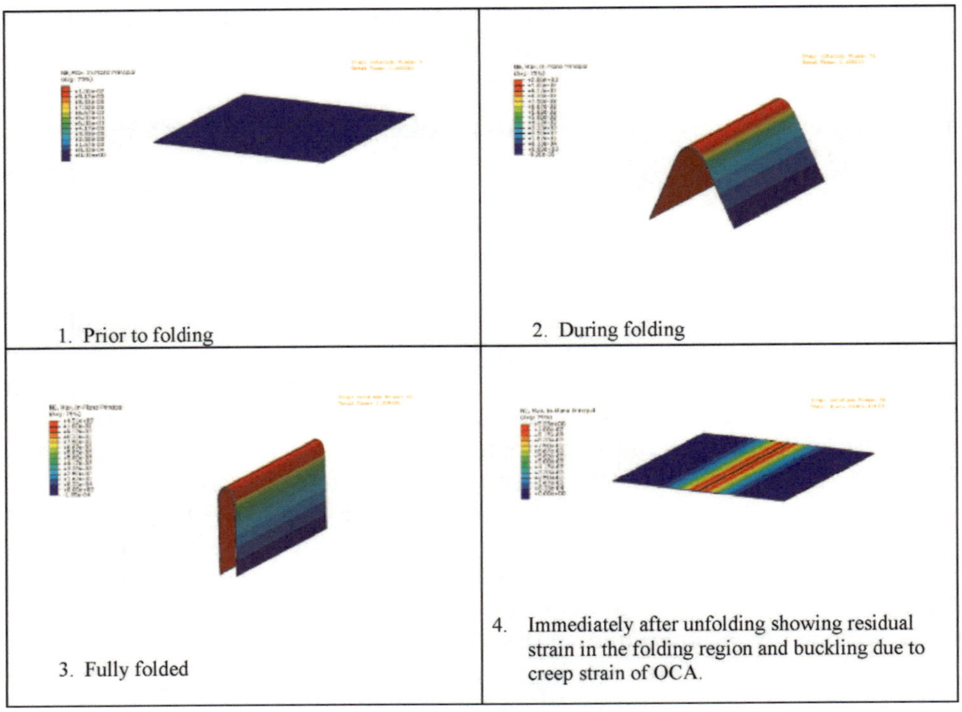

Figure 8.
Simulation results showing the 4 stages of folding. The film stack is folded towards its back layer in the form of the so-called "out-folding".

plane strain model meshed with the plane strain solid element, CPE4, was used. Six elements were used through the thickness for the OCA layer, and four elements through the thickness for each of the other functional layers.

4.4.1 Discussions of the simulation results

The following potential failure modes typical for foldable displays were predicted for various film stack layups by the simulation and are also observed in experiments:

- Failure due to tensile strain from bending exceeding the critical values for specific components/layers in the display film stack
- Buckling
- Debonding between layers of the film stack

Figure 9 shows the finite element analysis results of these failure modes. In the following sections, each will be discussed with emphasis on the effects of OCA's soft properties on the bending performance.

4.4.2 Fracture and failure due to tensile strain, multiple neutral planes due to large shear strain of OCA

For film stacks bonded by 3M™ CEF35 and 3M™ OCA 8211, the results show that the lower modulus of 3M™ CEF35 allows large shear slippage between layers leading to reduced bending stress in the film stack. The in-plane strain in the OLED

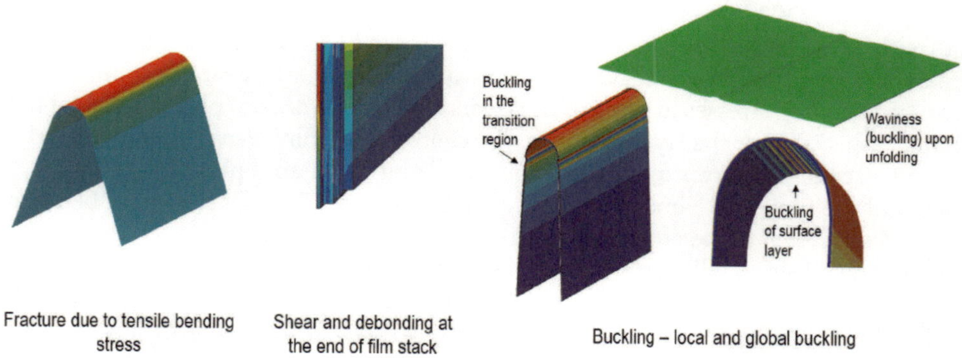

Figure 9.
The predicted potential failure modes of foldable display film stack.

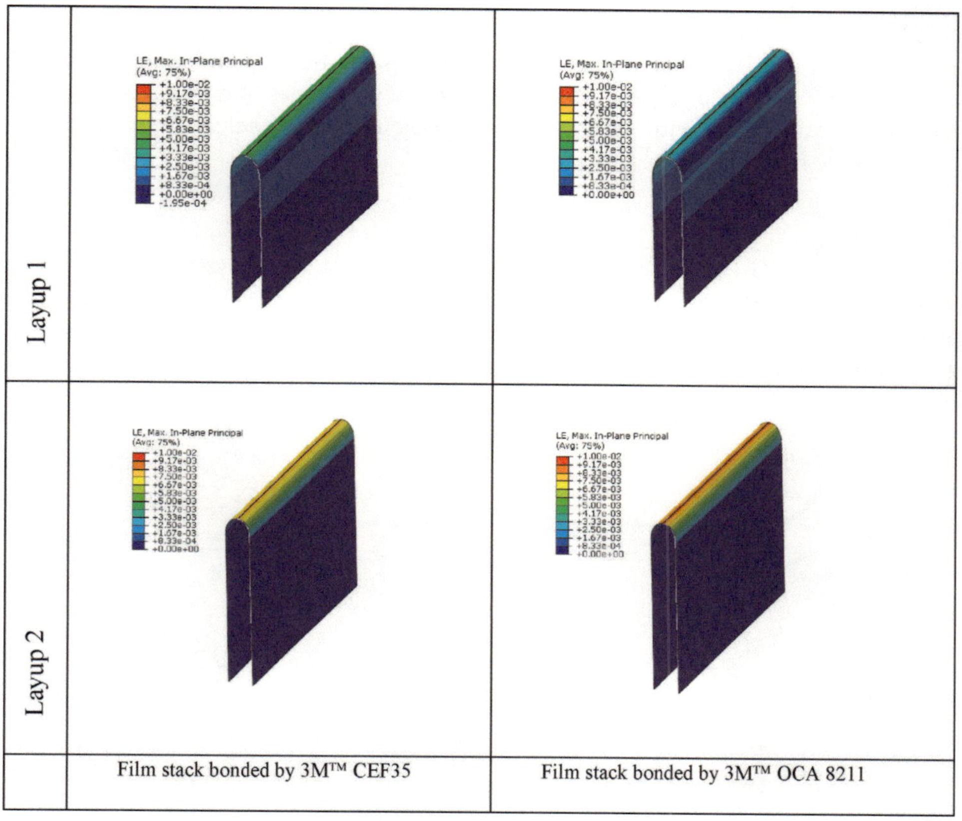

Figure 10.
Contour plots of in-plane bending strain in the OLED layer when the film stack is fully folded for the film stacks bonded by 3M OCAs (only the OLED layer is shown in this figure).

layer is shown in **Figure 10** for the two film stacks bonded by 3M™ CEF35 and 3M™ OCA8211. The maximum tensile strain occurs in the bend section at the bending symmetry plane after the film stack is completely folded. The maximum tensile strain ranged from 0.2% to 0.6% in the stacks bonded by 3M™ CEF35, and from 0.4% to 0.9% for stacks bonded by 3M™ OCA 8211. Depending on the critical strains of the OLED, ITO, or TFE, a less stiff OCA may make the difference between failure and no-failure of the display panel.

When the OCA's shear deformation between layers is sufficiently high to allow adjacent layers to slip over each other, each layer can bend without much constraint from the adjacent layers. This shear decoupling effect can result in much reduced bending stresses in some of the layers leading to a neutral plane in each of the layers, especially for the layers closer to the center of the film stack. In **Figure 11**, the bending stress (the stress in the direction parallel to the neutral plane of the film stack) of the film stack bonded by 3M™ CEF35 is shown. The plot shows the bending stress distribution at the cross-section of A-B, which is the location where the maximum bending stress will occur within the bend region of the film stack. For this 7-layer film stack of a total thickness of 0.475 mm (**Table 2**), the bending stress on each of the three OCA layers is negligible due to its low stiffness. The low stiffness of OCA, on the other hand, facilitates relatively high shear strain, in the range of 300–400% (**Figure 12**). This large shear strain between the layers allows each of the CWF, OLED, and back plate films to bend relatively independently from others resulting in a neutral plane in each of these functional layers, as indicated by the zero-bending stress location within each layer.

The ability of soft OCA to shear easily provides an effective design option to achieve multiple neutral planes in a film stack. This can be of great value since it can enable more freedom of choice regarding material properties and thicknesses of other functional layers without compromising their specific optical or electronic performance in a display.

4.4.3 Buckling

Under bending, a thin display film stack that is partially attached to a stiffer back plate is susceptible to buckling failure that leads to optical distortion and possible debonding, either within the film stack layers or from the back plates on which the

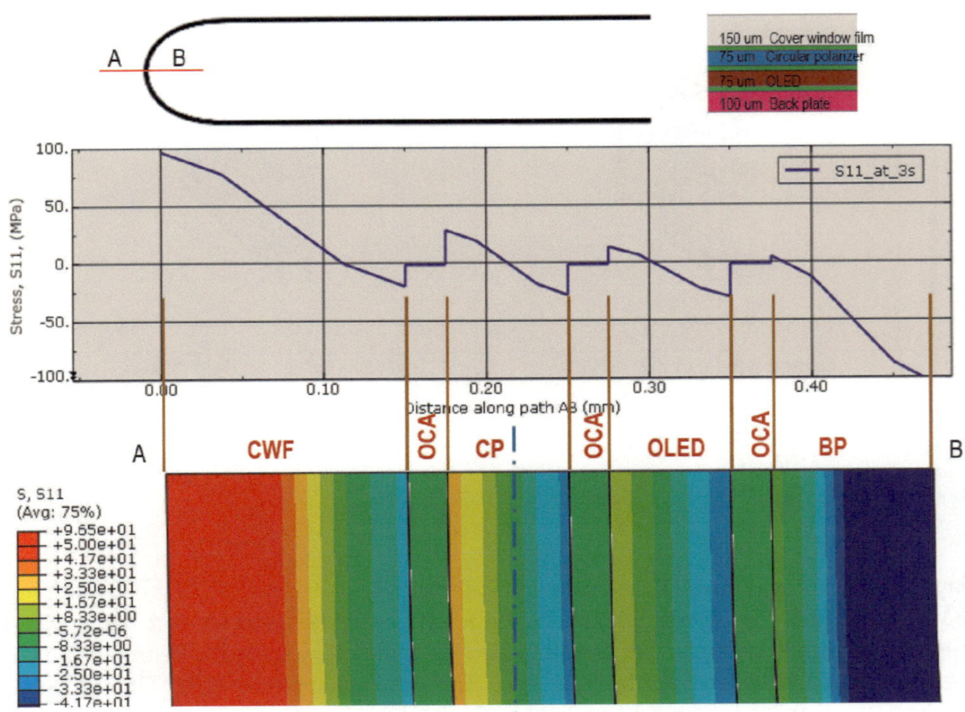

Figure 11.
Bending stress distribution at the cross-section A-B which is the plane of symmetry of the folded film stack.

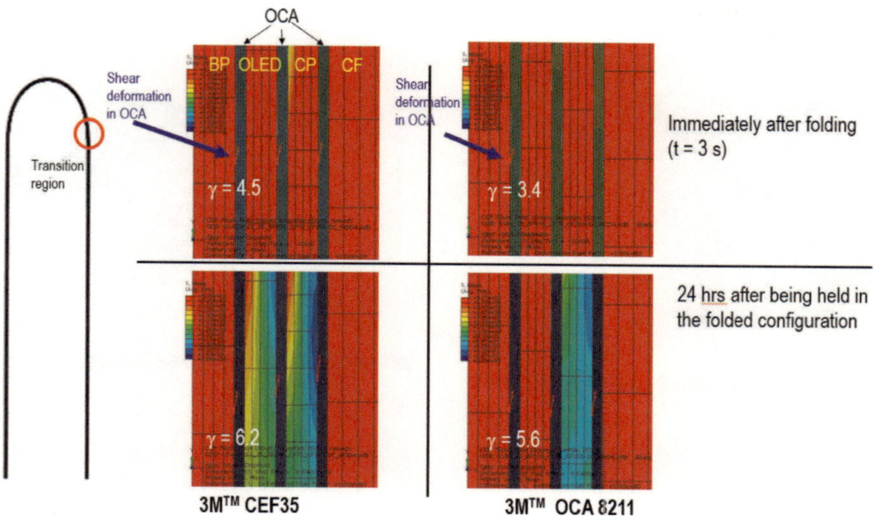

Figure 12.
Contour plots of shear stress in the transition region for the film stack (layup 1) showing the large shear deformation of the OCA layers.

film stack is attached. Buckling of the film stack can be further categorized into global buckling and local buckling types.

Global buckling typically gives the film stack a wavy appearance. It happens during unfolding after the film stack is held in the folded configuration for a period of time. In the folded configuration, shear creep in the OCA may occur which can be observed at the end of the film stack where each layer slips over the adjacent layer (**Figure 13**). This creep induced shear slip cannot be immediately recovered upon unfolding. During unfolding, the top layers of the film stack shown in **Figure 11** will then be under compression while the bottom layers will be in tension. Depending on the amount of the shear creep and the overall stiffness and thickness of the film stack, this may lead to instability and global buckling as shown in **Figure 14**. The more elastic behavior of 3M™ CEF35 limits the shear creep deformation and, in turn, leads to reduced buckling during unfolding.

Local buckling often appears as a fine and thin wrinkle or wrinkles on the surface on the compressive side of the folded film stack [36]. It occurs during folding but may still be visible after unfolding if the buckling is allowed to set in over a period of time or if the buckling causes debonding in the region.

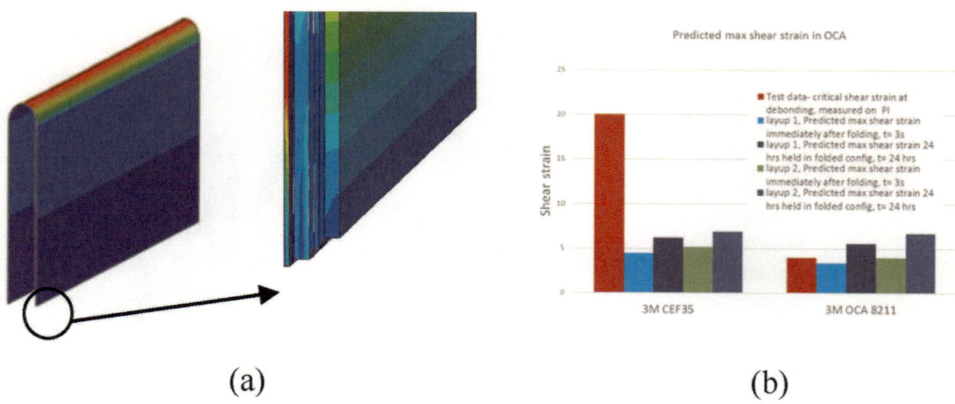

Figure 13.
Simulation results showing the shear deformation between layers (a), and the predicted maximum shear strain in OCA layer and the comparison with the bond strength of the OCA (b).

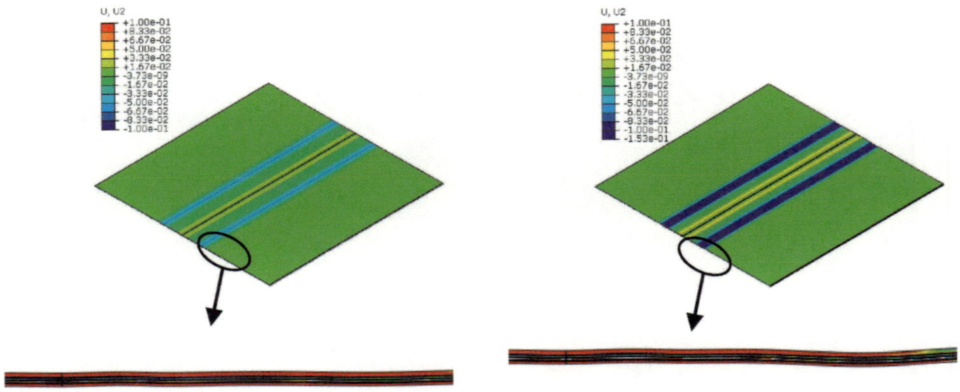

Film stack bonded by 3M™ CEF35 | Film stack bonded by 3M™ OCA 8211

Figure 14.
Displacement contour plots comparison of buckling of the film stack bonded by 3M™ CEF35 and 3M™ OCA 8211 immediately after unfolding showing the improved buckling resistance of 3M foldable OCA. The displacement, U2, is in the direction perpendicular to the film surface (unit: mm).

During folding, high compressive stress can develop in the bend region. Local buckling occurs only in the layers on or close to the surface of the film stack on the compressive side of the neutral plane. If the compressive stresses in these layers are not reduced through OCA's shear deformation, they may cause these layers to buckle or bulge out especially if any geometric off-axis imperfection is present in these layers. As shown in **Figure 15**, the buckling deformation will stretch the OCA that bonds the buckled layers to the remaining un-buckled layers resulting in tensile stress in OCA in the direction perpendicular to the bonding interface. Depending on the OCA bond strength, this tensile stress can cause delamination in this region.

Local buckling was not predicted in the film stacks bonded by 3M™ CEF35 or 3M™ OCA 8211 (the layup specified in **Table 2**). However, if the OCA modulus is reduced to 10% of that of 3M™ CEF35, the simulation predicted that local buckling could occur which is shown in **Figure 15** for a particular film stack. On the other hand, reducing the thickness of OCA to increase the constraint to buckling deformation can effectively improve the local buckling resistance as shown in **Figure 16**. Our studies have shown that reducing OCA thickness and increasing OCA stiffness can improve film stack's buckling resistance, while reducing OCA stiffness

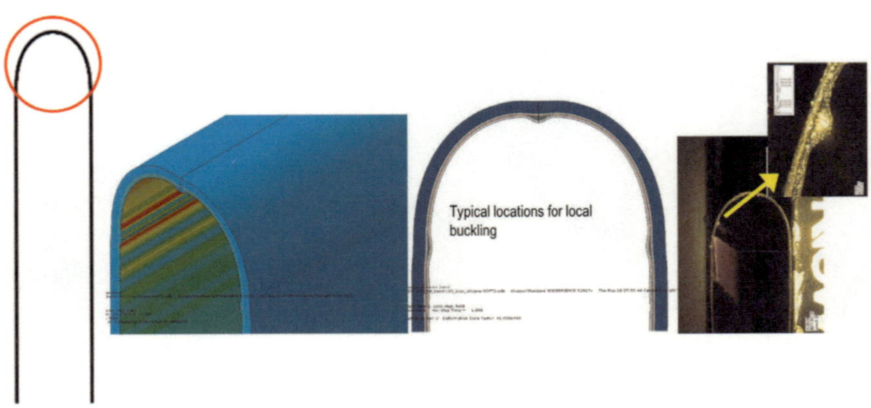

Figure 15.
Predicted local buckling on a film stack bonded with an OCA with a stiffness of 10% of that of 3M™ CEF35. The inserts are the photos of a local buckling on the actual film stack.

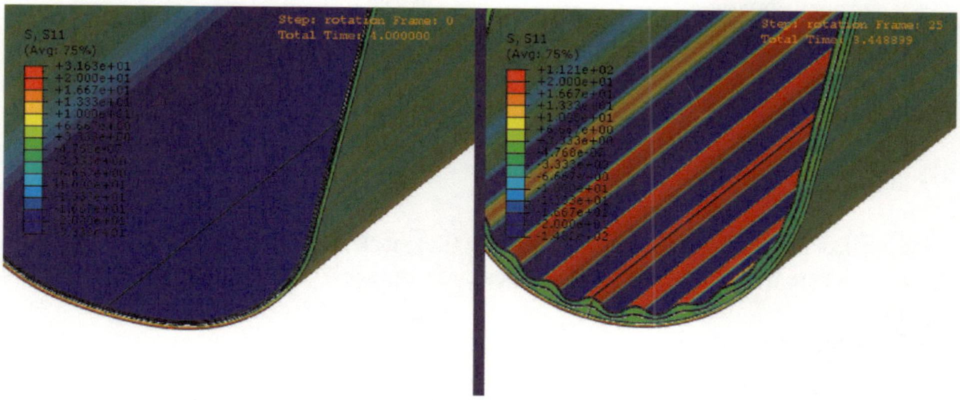

Figure 16.
Simulation results showing the effect of OCA thickness on local buckling. The film stacks shown here were bonded with 3M OCA of different thickness. The ratios of the surface layer thickness to the underlying OCA layer thickness is 2.0 for the layup on the left, and 1.0 for the layup on the right.

to facilitate shear slip is also critical in reducing the compressive stress that is the root cause of buckling. All these demonstrate that the design for buckling resistance is a balancing act and represents one of the most challenging aspects in film stack optimization. The competing requirements of stiffness to resist buckling and flexibility for folding, as well as, for minimizing the cause of buckling further highlights the importance of leveraging OCA's broader range of properties to achieve the optimal film stack performance.

4.4.4 Delamination and debonding

Another failure mode often observed in the foldable display is delamination or debonding, which is characterized by separation of layers either near the ends of the display panel, or within the bend region.

Typically, the debonding at the end of the film stack is due to shear strain in the OCA exceeding the OCA bond shear strength, while the delamination within the bend region is the result of high interlayer tensile stresses on OCA due to local buckling as described above.

In **Figure 12**, the shear stress of the film stack is shown for the transition region where the bent section meets the straight section. The corresponding shear strain values are indicated on the plots. **Figure 13** shows the shear stress at the end of the film stack. For both the film stacks bonded by 3M™ CEF35 and 3M™ OCA 8211, along the length of the film stack, the maximum shear strain occurs in the transition region. The higher the shear strain in the OCA, the more shear slip is allowed between layers, which reduces the bending stress in the layers. For the layers on the compressive side of the neutral plane, this means reduced compressive stress and the reduced possibility of local buckling of these layers. For the layups simulated in this work, the shear strain can be as high as 600% as shown in **Figure 12**. The softer 3M™ CEF35 allows not only large shear strain upon folding leading to improved local buckling resistance, it also can effectively resist debonding due to its high bond strength (**Figure 13**).

5. Summary

The simulation examples of folding OLED display film stacks discussed here demonstrate OCAs' abilities to reduce bending stresses and to create multiple neutral planes in a film stack. The simulation examples also demonstrate three potential failure modes for film stacks under dynamic folding and unfolding.

- Fracture due to high tensile bending stress
- Buckling
    - Local buckling of a few layers in the film stack and global buckling of the panel where the entire film stack buckles and assumes a wavy form
- Delamination or debonding

As shown in these simulation examples, these failures can be addressed through design and selection of appropriate OCAs. It is also evident that the folding performance of a display is a system-level response, where performance is governed by the properties of each layer in the stack layup as well as the display attachment methods and folding hinge design. The softest layers in the film stack, the OCAs, play a critical role in optimizing the film stack's folding performance when the options to alter the mechanical properties or thickness of other functional layers to improve overall stack flexibility are restricted by their designed functionalities. Design of display film stacks that utilize OCAs' unique characteristics can lead to increased robustness of integrated system solutions for foldable displays.

Conflict of interest

The authors declare no conflicts of interest.

Author details

Joel T. Abrahamson[1], Hollis Z. Beagi[1], Fay Salmon[2] and Christopher J. Campbell[1*]

1 3M Display Materials and Systems Division, Saint Paul, Minnesota, United States

2 3M Corporate Research Systems Lab, Saint Paul, Minnesota, United States

*Address all correspondence to: cjcampbell@mmm.com

References

[1] Which was the first mobile phone to market with a capacitive touchscreen? [Internet]. 2018. Available from: https://www.techspot.com/trivia/11-what-first-mobile-phone-market-capacitive-touchscreen/ [Accessed: 16 December 2018]

[2] Samsung Galaxy Round: The World's First Curved OLED Phone [Internet]. 2018. Available from: https://www.ign.com/articles/2013/10/09/samsung-galaxy-round-the-worlds-first-curved-oled-phone [Accessed: 16 December 2018]

[3] Samsung Galaxy Note Edge [Internet]. 2018. Available from: https://www.gsmarena.com/samsung_galaxy_note_edge-6631.php [Accessed: 16 December 2018]

[4] Israelachvili J. Intramolecular & Surface Forces. 3rd ed. London: Academic Press; 2011. 450 p. DOI: 10.1016/c2011-0-05119-0

[5] Pocius A. Adhesion and Adhesives Technology. 3rd ed. Carl Hanser: Munich; 2012. 386 p. DOI: 10.3139/9783446431775

[6] Kinloch A. Adhesion and Adhesives. Vol. 441p. London: Chapman and Hall; 1987. DOI: 10.1007/978-94-015-7764-9

[7] Dahlquist C. Tack. In: Adhesion: Fundamentals and Practice – A Report on an International Conference Held at the University of Nottingham, England; 20–22 September 1966; Nottingham. London: Ministry of Technology; 1969. p. 143

[8] Lakrout H, Sergot P, Creton C. Direct observation and fibrillation in a probe tack experiment on model acrylic pressure-sensitive-adhesives. The Journal of Adhesion. 1999;**69**:307-359. DOI: 10.1080/00218469908017233

[9] Campbell C. Optically clear adhesives. In: Chen J, Cranton W, Fihn M, editors. Handbook of Visual Display Technology. 2nd ed. Berlin: Springer; 2016. pp. 1501-1514. DOI: 10.1007/978-3-319-14346-0

[10] Everaerts A, Xia J. Cloud point-resistant adhesives and laminates. US Patent 8,361,632; 2013

[11] No DH, Kim JW, Kamine T, Hwang HD. Display Device Comprising Polarizing Layer. US Patent Application 20160093833; 2016

[12] Cho HS, Jin C, Kim EH, Yoo SH. Polarizer-free, high-contrast ratio organic light-emitting diodes utilizing microcavity structures and neutral-density filters. Journal of Information Display. 2014;**15**:195-199. DOI: 10.1080/15980316.2014.970238

[13] Gower M, Shanks R. Acrylic acid level and adhesive performance and peel master-curves of acrylic pressure-sensitive adhesives. Journal of Polymer Science Part B Polymer Physics. 2006;**44**:1237-1252. DOI: 10.1002/polb.20779

[14] Salmon F, Everaerts A, Campbell C, Pennington B, Erdogan-Haug B, Caldwell G. Modeling the mechanical performance of a foldable display panel bonded by 3M optically clear adhesives. SID Digest. 2017;**48**(1):938-941. DOI: 10.1002/sdtp.11796

[15] Royole Introduces 'FlexPai', the World's First Commercial Foldable Smartphone With a Fully Flexible Display, A Combination of Mobile Phone and Tablet [Internet]. 2018. Available from: https://www.royole.com/Dynamics?id=578 [Accessed: 08 January 2019]

[16] Galaxy Fold [Internet]. 2019. Available from: https://www.samsungmobilepress.com/galaxy-fold [Accessed: 20 March 2019]

[17] HUAWEI Mate X, 5G Smartphone, Foldable Design [Internet]. 2019. Available from: https://consumer.huawei.com/en/phones/mate-x/ [Accessed: 20 March 2019]

[18] LG OLED65R9PUA 65 Inch Class LG SIGNATURE OLED TV R9 [Internet] 2019. Available from: https://www.lg.com/us/tvs/lg-OLED65R9PUA-signature-oled-4k-tv [Accessed: 20 March 2019]

[19] Sun Z, Meng C, Ho J, Tseng K, Kwok H. Fabrication of broadband quarter wave plate by combination of two retardation films using coating technique. SID Symposium Digest of Technical Papers. 2018;**49**:222-224. DOI: 10.1002/sdtp.12687

[20] Zhou Y, Zhao Y, Reed JM, Gomez PM, Zou S. Efficient circular polarizer using a two-layer nanoparticle dimer array with designed chirality. Journal of Physical Chemistry C. 2018; **122**:12428-12433. DOI: 10.1021/acs.jpcc.8b02113

[21] Ma X-L, Chen X-C, Xiao L, Wang H-Y, Tan J-F, Song P, et al. P-126: A New Coatable Circular Polarizer for Anti-Reflection of Flexible AMOLED, SID Symposium Digest of Technical Papers, 2017; 48. DOI: 10.1002/sdtp.11998

[22] Liu Z, Mao X, Wang, M-HSI, Gu X, Shi S-m, Zhou W-F, et al. A Stack of Bendable Touch Sensor with Silver Nanowire for Flexible AMOLED Display Panel, SID Symposium Digest of Technical Papers, 2017; 48. DOI: 10.1002/sdtp.11788

[23] Fried A, Zhang XH, Abrahamson JT, Wang C, Luo J, Monson RJ, et al. Latest advances in silver nanowire based touch module reliability. In: 2015 IEEE 15th International Conference on Nanotechnology (IEEE-NANO). Rome; 2015. DOI: 10.1109/NANO.2015.7388838

[24] Letterier Y, Pinyol A, Gillieron D, Manson J-AE, Timmermans PHM, Bouten PCP, et al. Mechanical failure analysis of thin film transistor devices on steel and polyimide substrates for flexible display applications. Engineering Fracture Mechanics. 2010; **77**:660-670. DOI: 10.1016/j.engfracmech.2009.12.016

[25] Seo S-W, Jung E, Chae H, Seo SJ, Chung HK, Cho SM. Bending properties of organic–inorganic multilayer moisture barriers. Thin Solid Films. 2014;**550**:742-746

[26] Kim S, Kwon H-J, Lee HH, Shim H, Chun Y, Choi W, et al. Low-power flexible organic light-emitting diode display device. Advanced Materials. 2011;**23**:3511

[27] Jimbo Y, Tamatsukuri Y, Ito M, Yokoyama K, Hirakata Y, Yamazaki S. Reliability and mechanical durability tests of flexible OLED with ALD coating. Journal of the Society for Information Display. 2015;**23**:313

[28] Lee JH, Park JW, Lee TH, Shim GS, Kim HJ, Jung SY, et al. Variation of adhesion properties with molecular weight of cured and non-cured acrylic PSAs. In: Proceedings of the 39th Annual Meeting of the Adhesion Society. Vol. 151. 2016

[29] Beagi H, Corrigan T, Leatherdale C, Salmon F, Javid S, Harein M, et al. Folding test method development and differentiation of multi-layer concepts containing optically clear adhesive. In: Proceedings of the Adhesion Society; 25 February – 1 March 2018. San Diego; 2018

[30] Campbell CJ, Clapper J, Behling RE, Erdogan-Haug B, Beagi HZ, Abrahamson JT, et al. P-198: Optically clear adhesives enabling foldable and flexible OLED displays. In: Society for Information Display International Symposium Digest of Technical Papers.

Los Angeles; 2017. DOI: 10.1002/sdtp.12071

[31] International Electrotechnical Commission, IEC62715–6-2:2017

[32] Chiang CJ, Winscom C, Bull S, Monkmana A. Mechanical modeling of flexible OLED devices. Organic Electronics. 2009;**10**:1268-1274. DOI: 10.1016/j.orgel.2009.07.003

[33] Lee CC. Modeling and validation of mechanical stress in indium tin oxide layer integrated in highly flexible stacked thin films. Thin Solid Films. 2013;**544**:443-447. DOI: 10.1016/j.tsf.2013.02.084

[34] Jia YZ, Liu ZZ, Wu D, Chen JF, Meng H. Mechanical simulation of foldable AMOLED panel with a module structure. Organic Electronics. 2019;**65**: 185-192. DOI: 10.1016/j.orgel.2018.11.026

[35] Liu ZZ. Stress simulation of foldable OLED screen bending. Chinese Journal of Liquid Crystals and Displays. 2018; **33**(7):555-560. DOI: 10.3788/YJYXS20183307.0555

[36] Leterrier Y, Me'dico L, Demarco F, Manson J-AE, Betz U, Escola MF, et al. Mechanical integrity of transparent conductive oxide films for flexible polymer-based displays. Thin Solid Films. 2004;**460**:156-166. DOI: 10.1016/j.tsf.2004.01.052

Chapter 5

Picture Quality and Sound Quality of OLED TVs

Jang Jin Yoo, Dong Woo Kang, Jang Un Kwon, Sunho Park, Jeong Sub Hwang, Don Gyou Lee, Bu Yeol Lee and In Byeong Kang

Abstract

Unlike the past when cathode-ray tube (CRT) dominated display industry, many different types of flat panel displays (FPDs) are now leading the industry. Of these, organic light-emitting diode (OLED) display has recently become a next-generation display since this display is recognised as having advantages over other competing technologies in picture quality and form factor. With major attributes of picture quality considered, a series of evaluations based on objective measures was performed with an OLED TV compared to an LCD TV. OLED TV outperformed LCD TV 100 times in black, 20 times in colour contrast, 30% in dynamic range coverage, 50 times in local contrast and 20 times in viewing angle. In addition, sound quality of the OLED TV was assessed using both objective and subjective evaluation methods compared to conventional TV speakers since OLED panel speaker technology was recently commercialised. The OLED panel speaker showed better performance both in objective and subjective methods.

Keywords: OLED TV, OLED panel speaker, picture quality, sound quality, emissive display, HDR

1. Introduction

Although there are many different types of displays nowadays, displays can be mainly categorised as projection, off-screen, direct view and so on (**Figure 1**). Most of all, many FPDs of direct view displays have been commercialised very successfully. These FPDs can be also categorised in many ways, but the author will categorise them as an emissive display (or self-emissive display) and a non-emissive display in order to discuss picture quality of them because this kind of categorisation is highly related to picture quality.

Non-emissive displays need additional ambient light or illumination emitting light as a light source because colour and light level could not be expressed with the panel. In this category, there are electrophoretic display (EPD), electrochromic display (ECD) and so on, and the most successful one ever is liquid crystal display (LCD). For LCD, backlighting system is used as an illumination and each pixel of its panel controls colour and light level as an optical shutter [1].

On the other hand, an emissive display stands for the display which each pixel of its panel emits light with light level and colour controlled by itself [2]. For example,

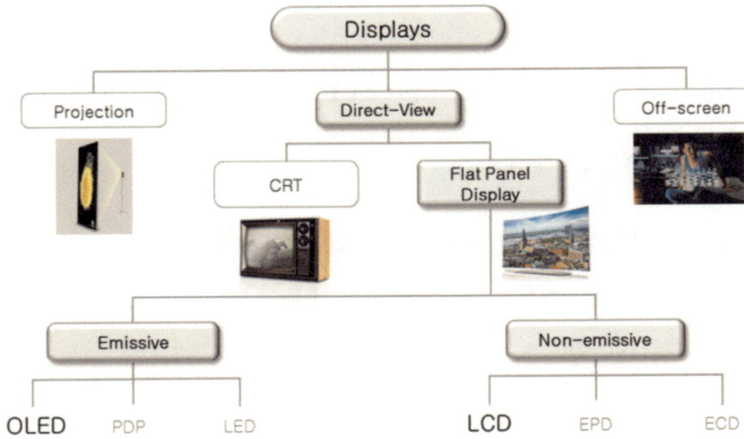

Figure 1.
Display category.

in this emissive display, there are plasma display panel (PDP), light-emitting diode (LED) and OLED which will be investigated in this chapter. OLED technology has been recently adopted for large TV displays and commercial displays as well as mobile displays [3].

This fundamental way of displaying causes the difference in picture quality. This chapter will introduce picture quality attributes and evaluation methods for high-end TV displays and assess the picture quality of OLED TV as an emissive display with the reference to LCD TV as a non-emissive display which is a current major TV display technology, and sound quality of OLED TV using OLED panel speaker will be also investigated.

2. Picture quality of OLED TVs

2.1 Evaluation method

In order to understand how to evaluate picture quality of OLED TVs, it is better to start with a well-known concept “image quality circle [4]”. This concept says that three steps—technology variables (display design factor), physical image parameters (physical measures) and customer perceptions (perceptual attributes)—are important for foreseeing customer image quality rating (picture quality that consumers consider), and these are organically linked and influence sequentially. Of these, physical measure and perceptual attribute, also called objective measure and subjective measure, respectively, should be measured together for an appropriate picture quality evaluation. In other words, these two measures should be linked for a perfect picture quality modelling [5].

In this study, only objective measure was considered. It will be also helpful to refer to a number of product reviews from various consumer reports in order to see a trend of subjective measure. Black, colour, viewing angle and HDR property were selected as the objective measures. Because recent high-end TVs have the identical specifications in resolution, colour bit depth, frame rate, etc., which were traditionally compared, other attributes causing differences in picture quality were chosen. The OLED TV used in this evaluation has 65 inches of diagonal size and 4 K ultra-high-definition (UHD) resolution. Also, 4 K UHD LCD TVs with the identical 65-inch size were compared to see a typical difference in picture quality between emissive display and non-emissive display. For the LCD TVs, two different backlighting types according to local dimming method were evaluated—edge-lit and direct-lit.

2.2 Evaluation result

2.2.1 Black

2.2.1.1 Pixel dimming vs. local dimming

First of all, the most fundamental performance of OLED TV which makes difference in picture quality is black level and pixel dimming. These performances could affect most of perceived picture quality attributes—colourfulness, brightness, sharpness, contrast and so on. That is why LCD manufacturers attempt to develop display technologies for lower black level with more number of dimming blocks.

For LCD, the technology which enables lower black level is called local dimming for backlight. The LCD TVs which were evaluated in this chapter have 12 dimming blocks for edge-lit backlight and 150 dimming blocks for direct-lit backlight (**Figure 2**). Since more dimming blocks secure lower black level and better picture quality, LCD manufacturers tend to increase the number of dimming blocks. Recent premium LCD TVs adopted over 500 dimming blocks. However, more dimming blocks mean more expensive cost for backlight and relevant electronic parts. Furthermore, direct-lit LCD for more dimming blocks should sacrifice thinner display thickness compared to edge-lit LCD showing better design.

However, OLED TV can inherently realise lowest black level because every single pixel can emit no light when black signal is applied. In terms of number of dimming blocks, OLED TV has more than 33 million blocks when 4 K UHD is considered (WRGB × 3840 × 2160). This is called pixel dimming because every sub-pixel (white, red, green and blue sub-pixels) is dimmable. Pixel dimming using emissive sub-pixels can reproduce peak highlight without halo artefact which normally appears in LCDs (**Figure 3**).

To increase brightness of peak highlight in LCDs, brightness of the LEDs in the dimming blocks corresponding to the peak highlight zone should be increased, and it will cause halo artefact around the peak highlight zone. On the contrary, if the brightness of the LEDs is decreased, the halo artefact could be reduced, but the peak highlight will get darker, and it will be difficult to reproduce the creative intent.

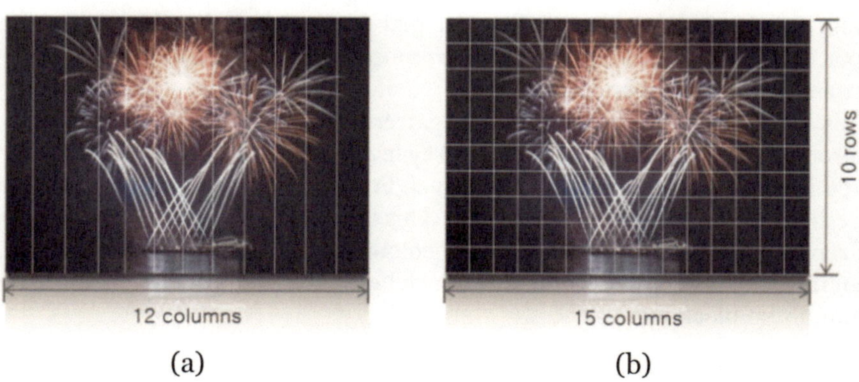

Figure 2.
LCD TVs according to local dimming method. (a) LCD TV with edge-lit backlight. (b) LCD TV with direct-lit backlight.

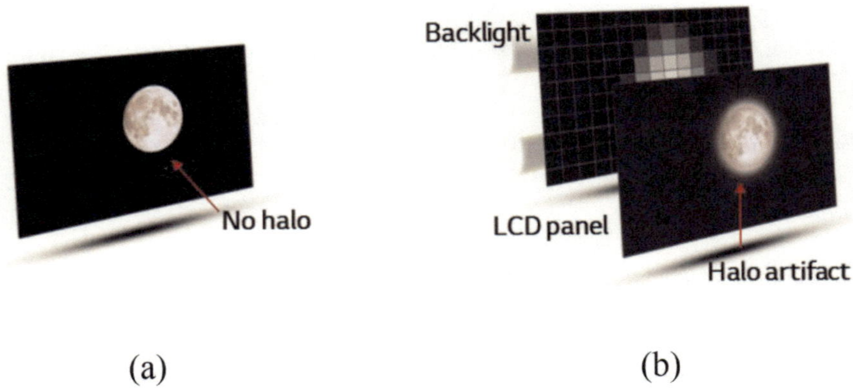

Figure 3.
Dimming performance comparison between OLED TV and LCD TV. (a) Pixel dimming (OLED TV) and (b) local dimming (LCD TV).

Figure 4.
Example of various colours around black signals in complex images and conventional black test pattern. (a) Black signals in complex images and (b) full-screen black [6].

2.2.1.2 Black level and black constancy

In this section, two objective measures for black performance described in the above section will be introduced. As can be seen in **Figure 4**, most complex images we face contain a variety of colour signals around black signals (or darker level signals). But conventional test patterns [6] in **Figure 4b** for measuring black performance of displays could not reflect the real situation like this. Thus a set of new test patterns as shown in **Figure 5** were proposed in the International Electrotechnical Commission (IEC) [7].

This set consists of five different test patterns with different grey levels of backgrounds (0, 255, 511, 767 and 1023 grey levels of overall 1024 grey levels) and a centred black box for measuring black level. Furthermore, black constancy value can be also evaluated with these patterns. This test shows how consistently the black level remains depending on neighboured colour signals. The black constancy C shown in Eq. (1) is defined as the difference between maximum black luminance and minimum black luminance.

$$C = \max[|L_{k,0} - L_{k,i}|] \tag{1}$$

where $L_{k,0}$ is the black luminance of the centre at the 0th grey background level; $L_{k,i}$ is the black luminance of the centre at the ith grey background level (i = 255, 511, 767 and 1023).

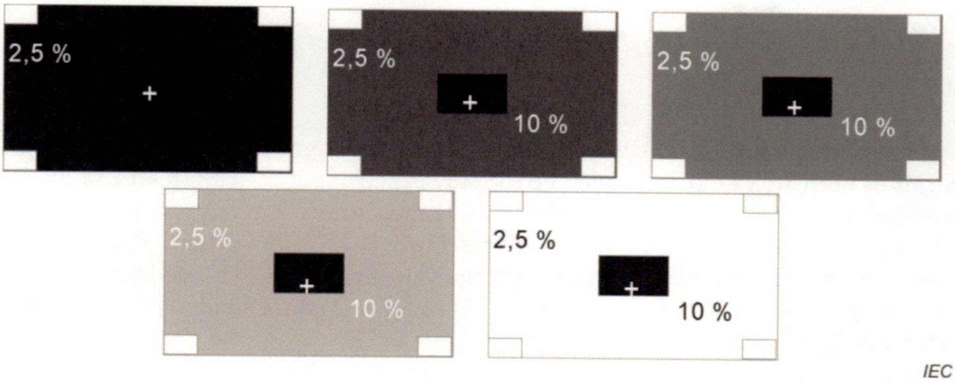

Figure 5.
Test patterns for black level and constancy measurement.

Black luminance (cd/m^2)	OLED TV	LCD TV (edge-lit)	LCD TV (direct-lit)
Max	8.3×10^{-4}	6.5×10^{-2}	8.2×10^{-2}
Min	5.6×10^{-6}	1.3×10^{-3}	1.7×10^{-3}
Ave	4.2×10^{-4}	$4.4 \times 10^{--2}$	5.1×10^{-2}
Constancy (Max-Min)	8.2×10^{-4}	6.4×10^{-2}	8.0×10^{-2}

Table 1.
Black luminance and black constancy of the OLED TV and the LCD TVs.

These black luminance data and black constancy values are summarised in **Table 1** and **Figure 6**. As can be seen, the black levels of LCD TVs are strongly influenced by the neighboured colour signals. However, OLED TV's black level remains comparatively constant showing 8.2×10^{-4} cd/m^2 black constancy. It was also found that direct-lit LCD showed bigger constancy than edge-lit LCD, which could be caused by higher brightness of direct-lit LCD.

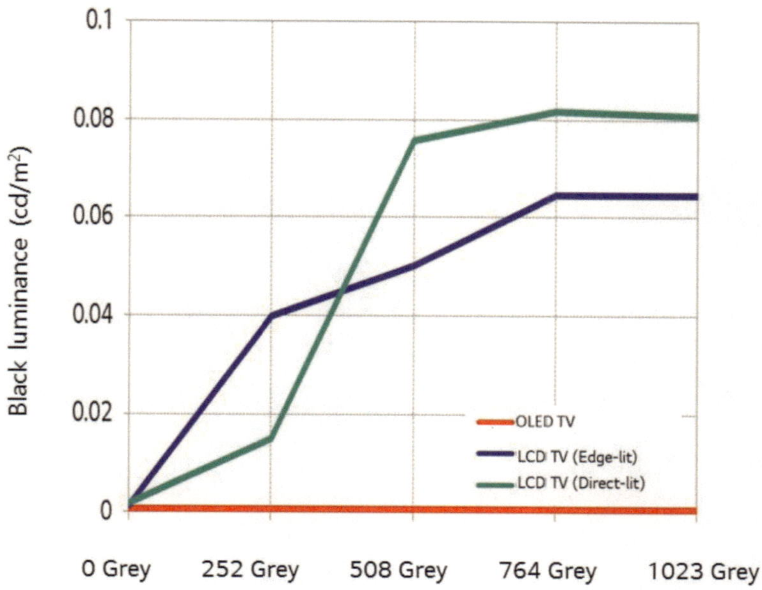

Figure 6.
Black luminance according to background grey levels.

2.2.2 Colour

2.2.2.1 Colour gamut

To investigate colour capability of OLED TV, colour gamut was evaluated in terms of both area ratio and coverage to standard colour spaces. Both area ratios with reference to BT.709 [8] and BT.2020 [9] show almost similar values for OLED TV and LCD TVs. Coverage values with reference to DCI-P3 [10] and BT.2020 also show similar for both displays. The coverage values for DCI-P3 exceed 95% in all the TVs compared. This indicates those TVs could reproduce similar colour capability for digital cinema.

Is it sufficient to explain the colour performance of those TVs with this evaluation? Since OLED TV is generally assessed more colourful at the similar colour gamut than LCD TVs [11], the reason will be discussed in the next section (**Figure 7**).

2.2.2.2 Colour contrast

Simultaneous contrast is one of well-known colour appearance phenomena which is easily perceived and experienced in our daily life [12]. **Figure 8** illustrates simultaneous contrast. The two identical grey patches presented on different backgrounds appear distinct. The darker background causes the grey patch to appear lighter, while the less dark background causes the white patch to appear less light. Although the centred colours have the identical luminance values, those two colours could appear different. This phenomenon can be found between TVs with different black levels.

To quantify this difference considering the relation between black level and colour, it is suggested to use colour contrast. The test pattern has black background and centred colour, and colour contrast is evaluated with the ratio of luminance of the colour to average luminance of the background (Eq. (2)) (**Figure 9**).

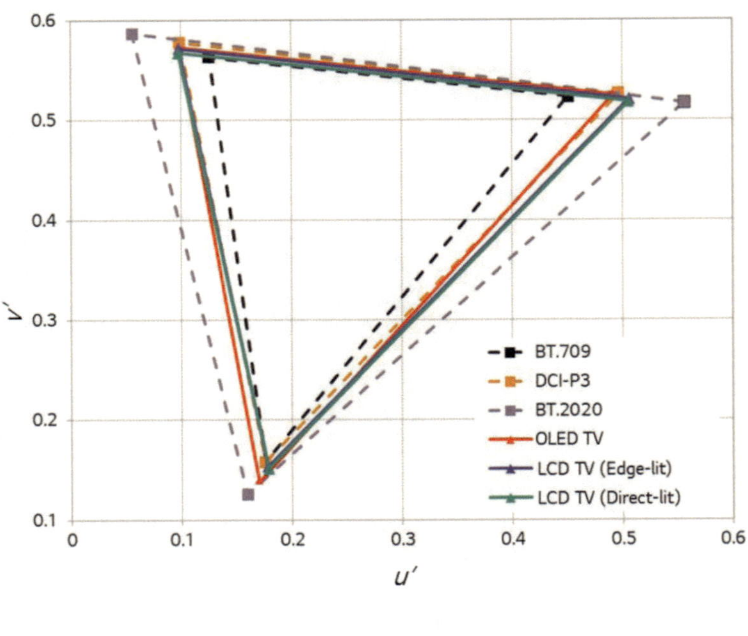

Figure 7.
*Colour gamut comparison of the OLED TV and the LCD TVs with reference to BT.709, DCI-p3 and BT.2020 colour spaces (**Table 2**).*

Figure 8.
Simultaneous contrast phenomenon perceived at the different dark backgrounds in luminance values.

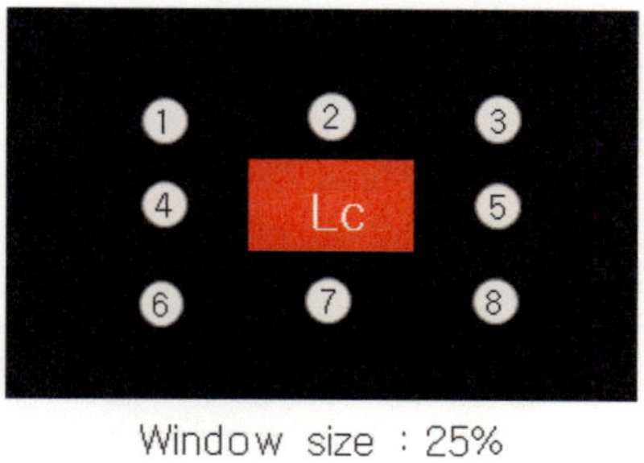

Figure 9.
Test pattern and measurement locations for colour contrast.

Colour gamut (in CIE1976)		OLED TV	LCD TV (edge-lit)	LCD TV (direct-lit)
Area ratio (%)	BT.709	129	128	128
	BT.2020	75	74	74
Coverage (%)	DCI-P3	99	97	96
	BT.2020	75	74	74

Table 2.
Colour gamut area ratio and coverage of the OLED TV and the LCD TVs.

$$\text{Colour contrast} = \frac{L_c}{L_k}, L_k = \frac{\sum_i^8 (L_i)}{8} \tag{2}$$

where L_c is the luminance of the centre and L_i is the black surround at the eight locations.

Colour contrast result was summarised in **Table 3**. OLED TV shows 60 times higher colour contrast than edge-lit LCD TV and around 20 times higher colour contrast than direct-lit LCD TV. This indicates that two TVs which have the same

Colour contrast	OLED TV	LCD TV (edge-lit)	LCD TV (direct-lit)
Red	120,000:1	1800:1	4700:1
Green	400,000:1	6500:1	17,000:1
Blue	64,000:1	790:1	2300:1

Table 3.
Colour contrast of the OLED TV and the LCD TVs.

colour gamut performance could show distinct perception depending on each black level, and this is one of the evidences in which black level affects many picture quality attributes such as brightness, contrast, sharpness, colourfulness, etc. as mentioned in the previous section.

2.2.3 High dynamic range (HDR)

HDR displays are recently becoming premium displays as more and more HDR contents are produced and they are rapidly distributed to the consumer market. HDR means a dynamic range higher than what is considered to be SDR (standard dynamic range) or LDR (low dynamic range) [13]. HDR generally offers better visual experience than SDR. In other words, HDR content could appear more similar to what human vision perceives. Especially, both bright details and dark details are expressed more naturally compared to SDR. Therefore, HDR TV should reproduce such highlight and dark zones more precisely.

Commercial HDR TV was first defined and announced by UHD Alliance (UHDA) in 2016 [14]. The UHDA announced their certification requirements for a HDR TV. According to the requirements, the HDR TV should have both a peak brightness of over 1000 cd/m^2 and a black level less than 0.05 cd/m^2 for LCD TV or a peak brightness of over 540 cd/m^2 and a black level less than 0.0005 cd/m^2 for OLED TV. These specifications are only the minimum requirement for HDR TV. After this logo programme was set up, many international standard bodies—IEC, International Committee for Display Metrology (ICDM), Video Electronics Standards Association (VESA), etc.—have made an effort to standardise metrologies or requirements. In this section, deeper consideration and evaluation of HDR properties for OLED TV will be made by a recent established international standard [7].

2.2.3.1 Dynamic range coverage

It is well known that human vision can perceive very wide range of brightness and darkness. There is some agreement that human vision can adapt to a value of 14 log units ranging from 10^{-6} to 10^8 cd/m^2 [15]. For a HDR content based on SMPTE ST.2084 [16], the code value ranges from o to 10^4 cd/m^2. If a reference dynamic range is given, a display's dynamic range coverage to the reference can be easily evaluated by measuring peak black luminance and peak luminance with the test patterns in **Figures 5** and **10**, respectively. That is, after measuring peak luminance using four patterns, the highest peak luminance is decided. Also the lowest black is decided with five black luminance patterns. Maximum dynamic range coverage (Eq. (3)) is obtained comparing the peak luminance value and black luminance value of reference [7]:

$$C_{dr.\max}[\%] = \frac{R_{dr.\max}}{\left(\log L_{w.ref} - \log L_{k.ref}\right)} \times 100 \tag{3}$$

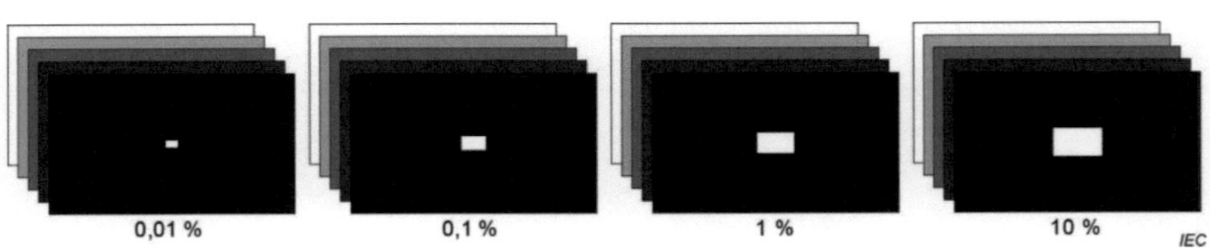

Figure 10.
Test patterns for peak luminance measurement.

$$R_{dr.\max} = \log L_{w.\max} - \log L_{k.\min} \quad (4)$$

where $L_{w,\max}$ is the maximum value of measured peak luminance and $L_{k,\min}$ is the minimum value of measured black luminance.

As a result, dynamic range coverages of OLED TV and LCD TVs are compared with reference to human vision and SMPTE ST.2084 content (**Table 4**). For the black level of SMPTE ST.2084, human vision's black was used instead of 0 cd/m^2. The OLED TV shows 61% of dynamic range coverage over human vision and 80% over SMPTE ST.2084 content, while LCD TVs show comparatively lower dynamic range coverage values. When the UHDA's required values are considered, two displays also show a big difference (OLED TV, 60%; LCD TV, 43%). From this result, it is clear that OLED TV still needs to achieve higher peak luminance in addition that the LCD TVs need to have lower black level and higher peak luminance in order to be a better HDR TV and cover higher dynamic range of human vision and HDR content. When, however, higher peak luminance is considered, visual fatigue needs to be considered as well [17].

2.2.3.2 Peak luminance and constancy

One of the great advantages of an ideal HDR TV is that it can express shiny peak highlights that cannot be experienced on SDR TVs. Therefore, peak luminance of HDR TV is very important to express such peak highlight zones properly. As shown in **Figure 11**, peak highlight appears in real images in a variety of sizes (**Figure 12**).

To see how large the peak highlights are distributed in the HDR content, all of the peak highlights of the content were detected, and statistical analysis was conducted [18, 19]. According to the analysis, it was found that the size of peak highlight, which appeared the most frequently, was 0.01% in area ratio and that the cumulative 90% of peak highlights was 0.2% in area ratio. This result indicates a very important point. That is, HDR displays should express small peak highlights within about 0.2% very well.

Dynamic range coverage (%)	UHDA (OLED)	UHDA (LCD)	OLED TV	LCD TV (edge-lit)	LCD TV (direct-lit)
Human vision	46	33	61	46	46
SMPTE ST.2084	60	43	80	60	60

Table 4.
Dynamic coverage of UHDA required displays, the OLED TV and the LCD TVs with reference to human vision and a HDR standard.

Figure 11.
Example of peak highlights in actual HDR images. (a) Peak size of 0.06% and (b) peak size of 1%.

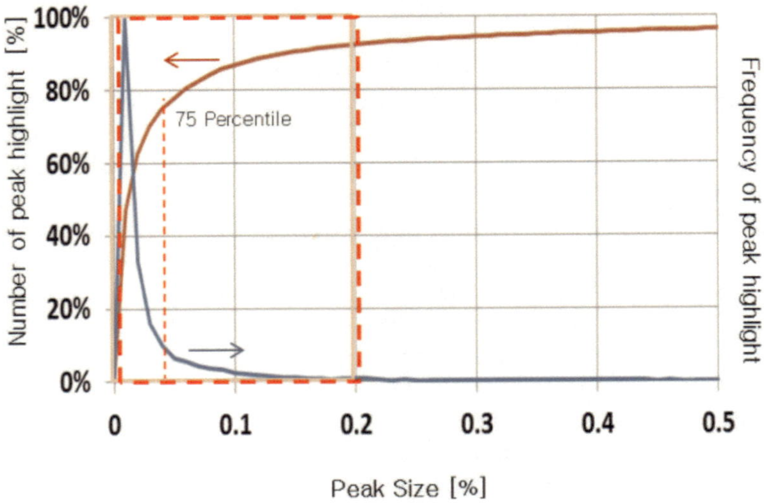

Figure 12.
Distribution of the size of peak highlights in actual images.

To evaluate these characteristics, peak luminance patterns introduced in **Figure 10** were used, and 20, 30, 40 and 100% patterns were additionally used. The peak luminance characteristics of the display can be well displayed according to peak highlight size. The results of the measurements are detailed in **Figure 13**. LCD TVs had higher luminance values than OLED TVs at peak highlight, which is larger than 1% size. On the other hand, OLED TV showed almost constant peak luminance values at smaller peak highlight size than 10%.

As mentioned about the UHDA requirement in the preceding section, the peak luminance of LCDs 1000 cd/m^2 is taken equivalent to 540 cd/m^2 of OLED TV because of the high contrast characteristic of OLED TVs. Given this fact, OLED TV can be said to have consistently high peak luminance not only at 10% of the peak highlight but also at smaller than 0.01% size. However, only from 1 to 10% size, LCD TVs have high peak luminance of 1000 cd/m^2. As noted in the previous

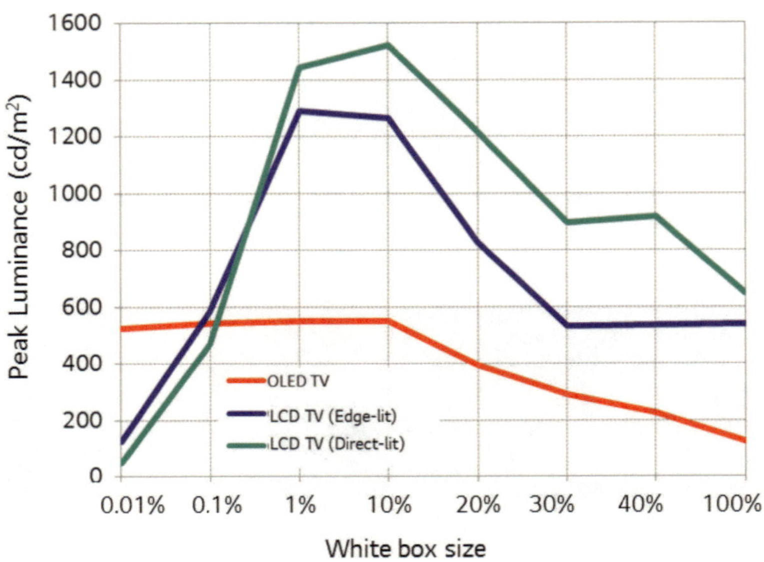

Figure 13.
Peak luminance of the OLED TV and the LCD TVs according to the size of the white boxes.

Size (%)	OLED TV	LCD TV (edge-lit)	LCD TV (direct-lit)
0.01	526	123	49
0.1	544	587	470
1	549	1288	1444
10	548	1263	1523

Table 5.
Peak luminance according to the size of white box.

section, it can be said that OLED TV is very advantageous in expressing peak highlight because peak luminance is very important in smaller peak highlight size than 0.2% (**Table 5**).

In addition, the results of measuring peak luminance were plotted by varying the grey level around the white box in the measurement pattern (**Figure 14**). As can be seen, the OLED TV implements the maximum peak luminance regardless of the surrounding signal compared to the LCD TV, while the LCD TV implements the maximum peak luminance within a relatively small range. In the actual video, where various signals are inputted, an OLED TV can show consistently bright peak luminance.

2.2.3.3 Local contrast

Sequential contrast measurement methods, which used to be mainly used, measure the black luminance and peak luminance values of the display in different patterns and then determine the contrast ratio by the ratio of those values. However, with this value, it is difficult to represent the local contrast characteristic frequently encountered in actual content because dark and bright parts usually appear at the same time in real life (**Figure 15**).

In order to measure another important characteristic of HDR, a local contrast, a measurement method was proposed using the pattern shown in **Figure 16** and Eq. (5) [7]. Measurements showed that OLED TV showed local contrast values 50 times larger than LCD TVs (**Table 6**):

$$CR_L = \frac{8L_C}{L_1 + L_2 + L_3 + L_4 + L_6 + L_7 + L_8 + L_9} \tag{5}$$

where L_c is the luminance of the centre; L_i is the black surround at the eight locations.

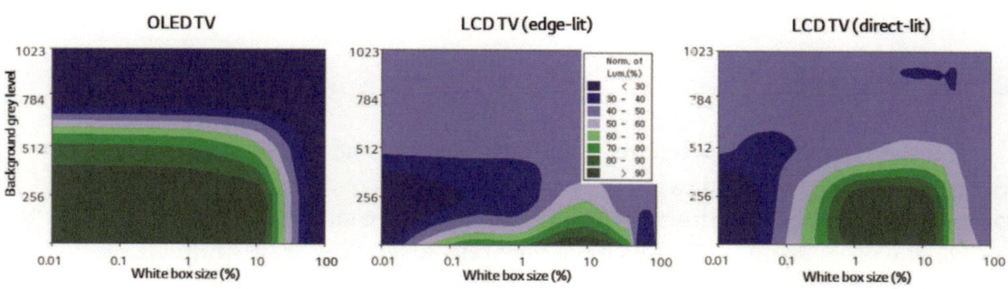

Figure 14.
Peak luminance of the OLED TV and the LCD TVs influenced by grey level of background.

Figure 15.
Example of the images that local contrast measurement is important.

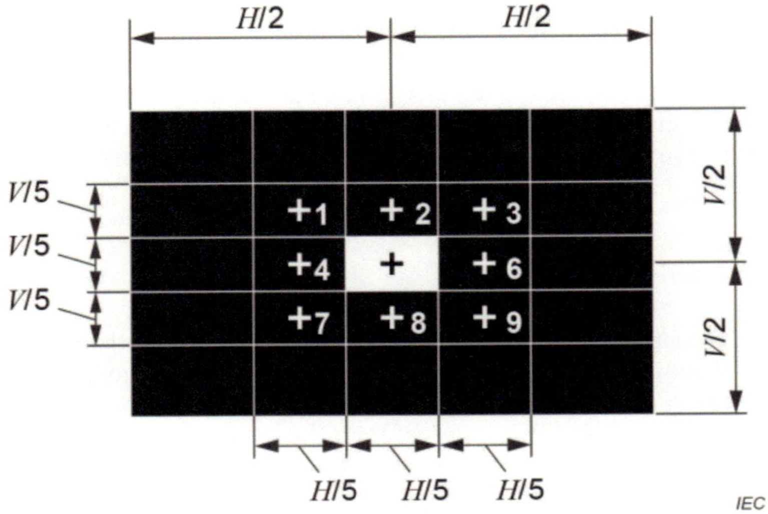

Figure 16.
Test pattern for local contrast.

	OLED TV	LCD TV (edge-lit)	LCD TV (direct-lit)
White luminance	544	1280	1420
Ave. black luminance	0.0005	0.0936	0.0729
Local contrast	1088K:1	13.6K:1	19.5K:1

Table 6.
Local contrast of the OLED TV and the LCD TVs (cd/m^2).

2.2.4 Viewing angle

Another big difference between emissive display and non-emissive display is the performance of the viewing angle. OLED TV has relatively constant optical properties in the entire direction as light emits from each pixel of OLED panel.

On the other hand, LCD TVs have different displaying ways from OLED TV because the light emitted from backlight passes through liquid crystal layer which is optically anisotropic. When the light enters the pixelated liquid crystal layer, it is polarised. The polarised light goes through phase retardation because of optical anisotropic property of liquid crystal layer. The phase retardation depends strongly on the incident angle. Therefore, picture quality is influenced by the viewing angle. That is, colour, gamma, luminance and black characteristics will vary depending on the viewing angle (**Figure 17**).

For colour shift criteria ($\Delta u'v' < 0.02$), OLED TV is 120°, while LCD TV is 92°. For half-luminance angle where luminance is reduced by half to the front luminance, OLED TV is measured 132°, while LCD TV is 62°. In addition, for gamma shift ratio (GSR), which measures the extent of gamma variation at off-axis to the normal direction, the OLED TV showed 1% variation, while LCD TVs showed around 20% shift (**Table 7**). This numerical difference causes colour distortion as shown in **Figure 18**.

Next, black performance and HDR properties that were examined in the previous section were compared at 45° viewing angle. The result is summarised in **Table 8**. While the OLED TV showed the identical black luminance value to the value at the normal direction, the black luminance values of LCD TVs have increased by four times and six times, respectively. For peak luminance, although the OLED TV maintained at 94% at off-axis compared to normal direction, the LCD TVs showed around 50% or less of peak luminance compared to normal direction.

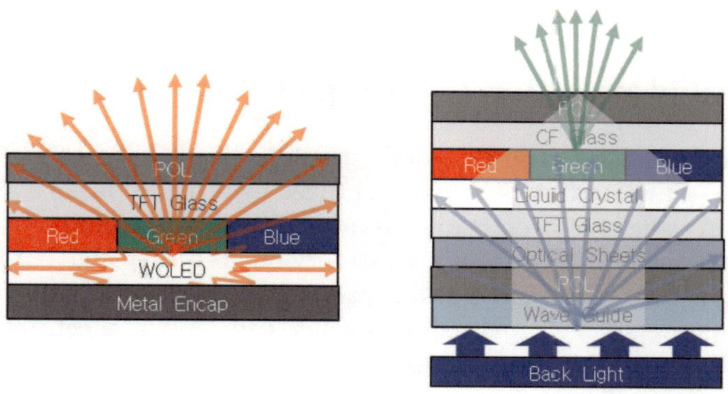

Figure 17.
Conceptual diagrams of light path of OLED TV and LCD TV. (a) OLED TV and (b) LCD TV.

Gamma shift ratio	OLED TV	LCD TV (edge-lit)	LCD TV (direct-lit)
0° vs. 45° (%)	1.0	19.6	23.0

Table 7.
Gamma shift ratio of the OLED TV and the LCD TVs.

Figure 18.
Picture quality change at the off-axis of the OLED TV and the LCD TVs.

Picture quality attributes	OLED TV			LCD TV (edge-lit)			LCD TV (direct-lit)		
	0°		45°	0°		45°	0°		45°
Black luminance (cd/m^2)	5.6E−6	100% →	5.6E−6	1.3E−3	640% →	8.3E−3	1.7E−3	440% →	7.4E−3
Peak luminance (cd/m^2)	548	94% →	516	1263	34% →	430	1523	49% →	749
Dynamic range coverage (%)	80	100% →	80	60	78% →	47	60	83% →	50
Local contrast (:1)	1.1M	91% →	1.0M	13K	7.7% →	1K	19K	11% →	2K

Table 8.
Black and HDR properties of viewing angle of the OLED TV and the LCD TVs.

In addition, the dynamic range coverage and local contrast of the OLED TV were also relatively maintained compared to the normal direction, whereas LCD TVs have deteriorated at the 45° viewing angles. In summary, the OLED TV generally maintains picture quality at the off-axis direction.

3. Sound quality of the OLED TV using OLED panel speaker

OLED panel speaker has been recently developed by LG Display, which vibrates the thin OLED panel itself with actuator or piezo materials to make a sound on the viewing screen to deliver directly to consumers. The sound reproduced on the centre of panel not only increases sound quality but also keeps the borderless and thin TV design. The OLED panel speaker can maximise vibration transfer efficiency compared to non-emissive panel speakers which have optical sheet, reflector and light guide plate in addition to the panel.

At the Consumer Electronics Show (CES) in 2017, LG Display also demonstrated a technology called Crystal Sound OLED (CSO), which has surpassed the limitations of conventional TVs using indirect reflected sound. This technology makes consumers feel like they hear the sound coming directly from the OLED TV screen, not the reflected sound coming from a separate speaker built into the TV.

By combining the technology to transmit its own sound with the OLED technology, only the advantages of both external and internal speakers applied to TV have been secured with the differentiated CSO technology.

In this section, the level of sound quality was confirmed through objective and subjective evaluation based on the difference between the position and radiation direction of TV speaker. Moreover the sound quality was evaluated between CSO and conventional TVs [20].

3.1 Objective measures

3.1.1 Preliminary experiment by speaker position

Prior to evaluating the speaker performance of a TV product, the sound quality of one channel was quantitatively measured by viewer's location as shown in **Figure 19** with a 65-inch TV product, whose speaker is located at the bottom and has a structure that radiates sound forward.

For this experiment, the x axis in **Figure 20** represents the frequency and the y axis the frequency response, which is the sound pressure level. The experiment was

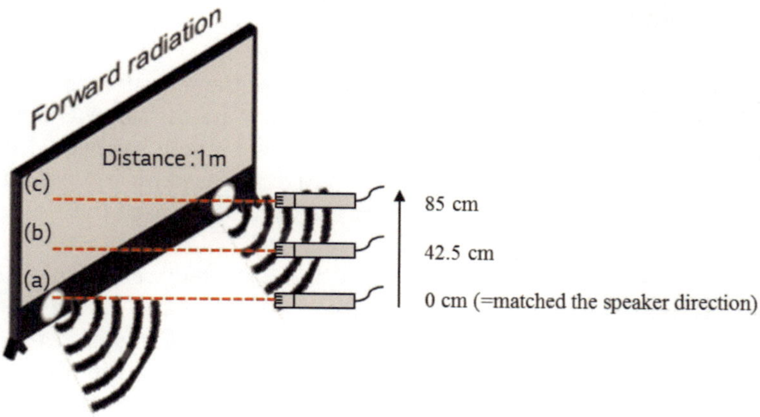

Figure 19.
Forward radiation structure, test positions and the distance for the TV sample.

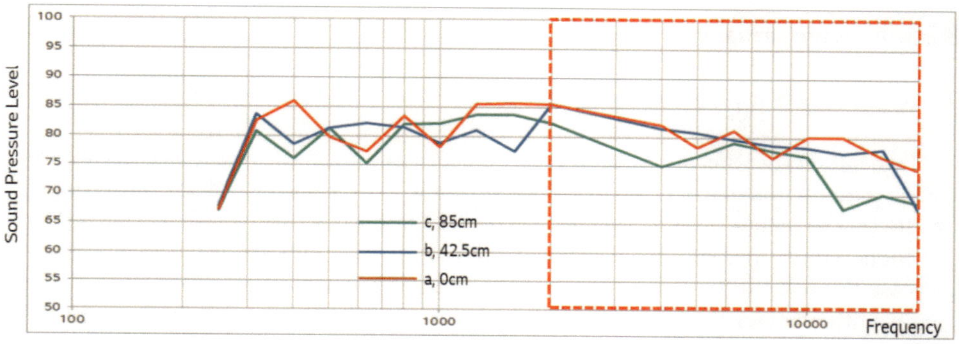

Figure 20.
Frequency response results by speaker height.

conducted in an anechoic chamber to rule out the space characteristics, and the frequency band has been limited from 250 Hz to 20 kHz to reflect the space characteristics of the real anechoic chamber. Also, a stepped signal with 1/6 octave of 20 Hz to 20 kHz has been used as a test signal, and the measurement distance is 1 metre, and the measurement position is shown in **Figure 19**.

"(a)" is located on the principal axis of the speaker, and "(b)" and "(c)" indicate positions that are increasingly distant from the speaker. As can be seen in **Figure 20**, the measured results show that the sound pressure level in the band larger than 3 kHz was kept flat without decreasing at "(a)". On the other hand, the sound pressure level was drastically reduced in the band larger than 15 kHz at "(b)". At "(c)", the sound pressure level was reduced more in the band from 2 to 10 kHz when compared to "(a)" and "(b)", and the sensitivity decreased by about 10 dB in the band larger than 10 kHz. This result shows that it is difficult for the consumer to experience the original sound quality of the speaker when performance degradation occurs in the high frequency band with increasingly receding viewing position and speaker height position in terms of TV user experience.

3.1.2 Sound quality evaluation of OLED panel speaker

Thus, the frequency responses of the following three products were measured to be compared—OLED panel speaker (A), a product with a speaker at the bottom that radiates sound forward (B) and a product with a speaker at the bottom that radiates sound downward (C).

The speaker position and the radiation structure of the measurement samples are shown in **Figure 21**. The conditions of the experiment are the same as the previous one. Also, in the case of the measurement position, the screen has been divided into the speaker channels in order to simulate the situation based on audiovisual standards in accordance with the scenario where the consumers used to be measured, respectively, by the centre of the divided screen.

The measurement results for products A, B and C are shown in **Figure 22**. As can be seen in the graph, the sound pressure level of the OLED panel speaker tends to be flat in the band from 2 to 20 kHz. On the other hand, B and C products tend to have a lower sound pressure level in the band larger than 8 kHz when compared to product A. This graph also shows that product A with an OLED panel speaker has the advantage of the physical location of the speaker, which allowed the characteristics of the frequency response to remain flat up to the high frequency band at the measuring position for viewing.

3.2 Subjective measures

3.2.1 Preliminary experiment by speaker position

Two subjective measures have been made to compare the sound quality of an OLED panel speaker with that of a conventional TV speaker. Prior to comparing the products, the sound quality by the speaker's position has been compared for TV products whose speakers are located at the bottom and radiate sound forward to increase the reliability of this experiment. That is, the speaker is located at the ear height of the viewer (condition A), 30 cm below the viewer's ear height (condition B)

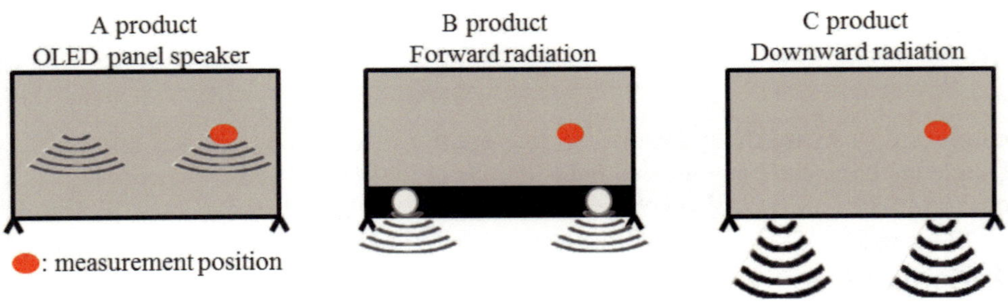

Figure 21.
Speaker position and radiation structure for the TV samples.

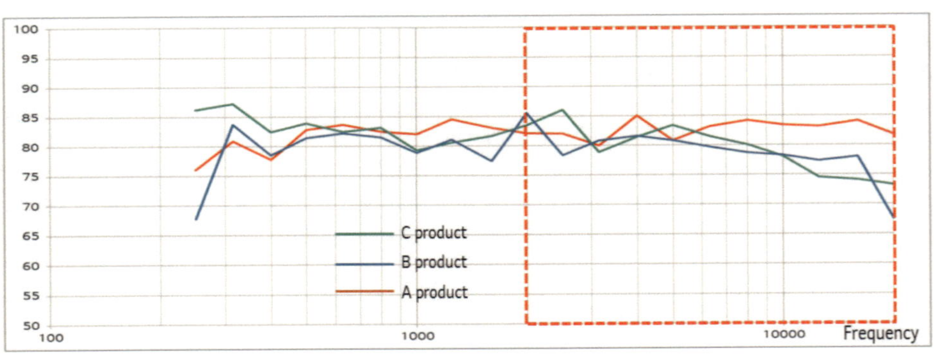

Figure 22.
Frequency response results. (A) OLED panel speaker, (B) forward radiation and (C) downward radiation.

and 60 cm below the viewer's ear height (condition C). **Figure 23** shows that a player called foobar2000 was used on the PC, the PC and TV were connected with HDMI, the height of the assessment samples was designed to be adjusted quickly with an automated electric table and a 9-point scale was used for assessment. Test sound loudness was 63 dB (C), and source type was FLAC, 44.1 kHz, and 16 bit. The items in **Table 9** based on EBU document Tech.3286 have been adopted for the assessment [21]. A, B and C conditions were applied rotationally to eliminate the influence of the experimental procedure [22].

As shown in **Figure 24**, condition A (where the speaker is located at the ear height of the viewer) has had higher scores in preference by about 1.5 points, in sound location accuracy and timbre by about 2 points and in sound balance by about 1.5 points when compared to condition C (where the speaker is located 60 cm below the viewer's ear height). However, the correlation r for spatial impression was 0.26, which indicates that there is no significant difference within the range of 60 cm downward. As a result, it can be deduced that in an environment where you watch TV, the increasing difference between the height of the TV speaker and that of your ears negatively affects the sound quality factor of preference, sound location accuracy, timbre and sound balance. This experiment has improved the reliability of the subjective assessment of the sound quality of three TV speakers with different speaker drive units.

3.2.2 Sound quality evaluation of OLED panel speaker

Next, an evaluation was carried out on the following three TV products for subjective assessment: product A whose speaker is located at the bottom and radiates sound forward, product B with an OLED panel speaker and product C whose speaker is located at the bottom and radiates sound downward. The experimental

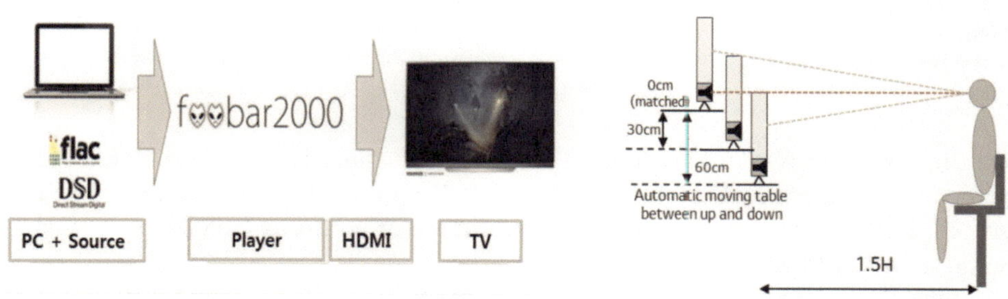

Figure 23.
Test conditions for subjective assessment according to the speaker height.

Parameter	Check point	Test programme
Spatial impression	Spatial size, balance	Classic, POP
Sound location accuracy	Location accuracy	Classic
Transparency	Sound source definition	Classic, piano solo
Sound balance	Dynamic range	Classic
Timbre	Sound colour, frequency	Guitar, POP
Overall preference	Preference	All

Table 9.
Test parameter and programme for subjective assessment.

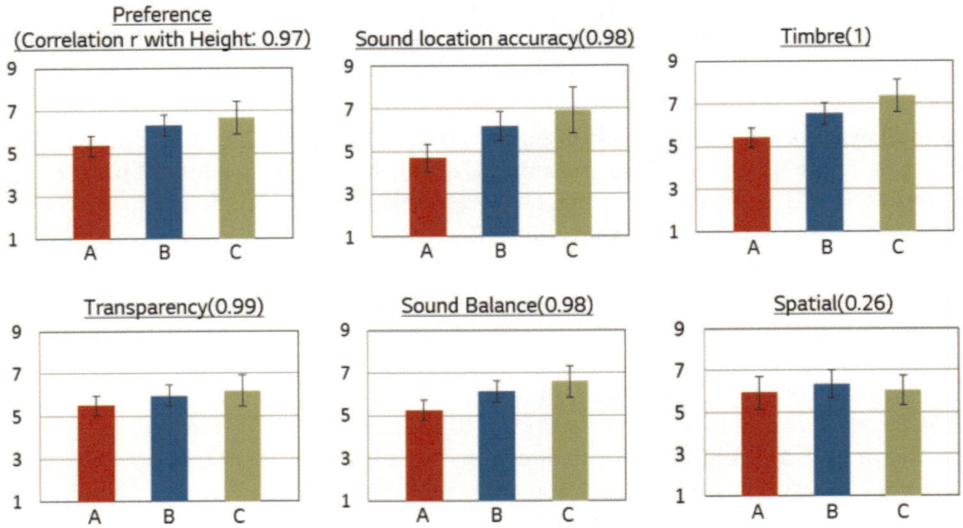

Figure 24.
Results of the subjective assessments by speaker height.

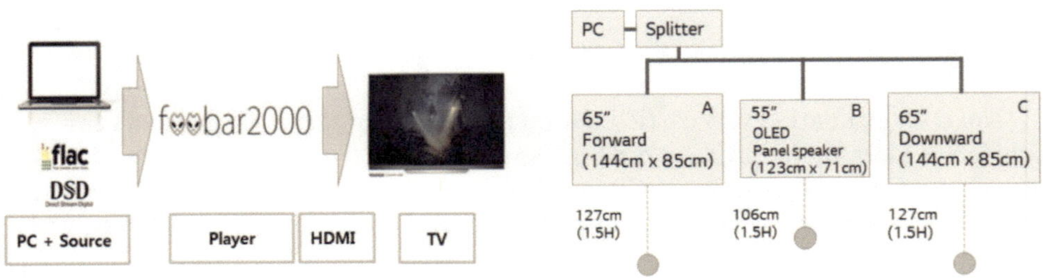

Figure 25.
Test conditions for subjective assessment according to speaker radiation structure of the TV samples.

conditions are shown in **Figure 25**. The viewing distance has been set to 1.5H (H referring to the vertical length of the TV panel) depending on the size of each TV screen. Both the assessment items and the source programme have been set the same as in **Table 9**.

According to **Figure 26**, product B with an OLED panel speaker has received the highest score for sound location accuracy, timbre and transparency. Moreover, a higher score was received for overall preference in descending order of B, A and C, which is the order where the ears and the screen coincide with each other on the horizontal axis.

For spatial impression, however, there was no significant difference between the three products just as in the previous experiment. As a result, it can be deduced that in an environment where you watch TV, the increasing difference between the height of the TV speaker and that of your ears negatively affects the sound quality factor of preference, sound location accuracy, timbre and sound balance.

Based on the above two experiments, it has been shown that observers would have the best sound quality and preference when the height of their ears matches that at which the speaker radiates sound, at a distance of less than 60 cm, the speaker height does not significantly affect spatial impression and at the moment, most speakers applied to TVs are located at the bottom and radiate sound downward to have poor sound quality for timbre, transparency and preference.

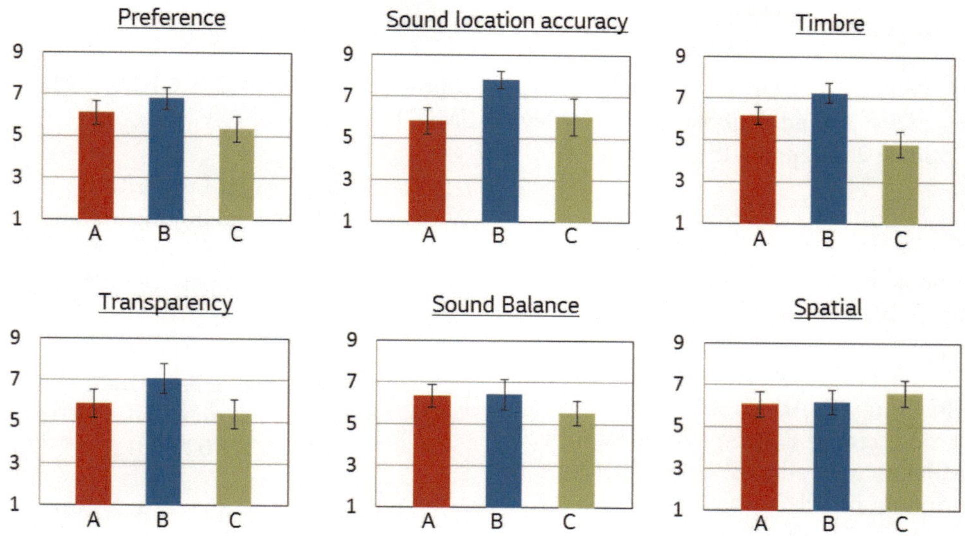

Figure 26.
Results of the subjective assessments according to speaker radiation structure of the TV samples.

4. Conclusions

This chapter dealt with picture quality of OLED TV and sound quality of the OLED TV using the OLED panel speaker. Reasonable picture quality attributes for the evaluation were introduced, and this evaluation was performed using the measurement methods based on several international standards. The OLED TV evaluated here showed better performance in black, colour, HDR and viewing angle properties compared to high-end LCD TVs. It was also proved that the fundamental reason of the difference is due to the emissive display which emits light at every single pixel.

Sound quality of the OLED panel speaker was then evaluated in both objective and subjective ways. It showed better quality particularly in higher frequency and was evaluated better in most subjective assessment attributes when compared to the conventional TV speakers, which indicates this technology is one of the good solutions for on-screen sound.

Author details

Jang Jin Yoo*, Dong Woo Kang, Jang Un Kwon, Sunho Park, Jeong Sub Hwang, Don Gyou Lee, Bu Yeol Lee and In Byeong Kang
LG Display, Seoul, Republic of Korea

*Address all correspondence to: 1yoo@lgdisplay.com

References

[1] Yeh P, Gu C. Optics of Liquid Crystal Displays. 2nd ed. Hoboken: John Wiley & Sons; 2010. pp. 1-15

[2] Tsujimura T. OLED Displays: Fundamentals and Applications. 2nd ed. Hoboken: John Wiley & Sons; 2017. pp. 5-10. DOI: 10.1002/9781118173053

[3] Han CW, Choi H, Ha C, Shin H, Choi HC, Kang IB. Advanced technologies for large-sized OLED display. In: Ravariu C, Mihaiescu D, editors. Green Electronics. London: IntechOpen; 2018. pp. 33-34. DOI: 10.5772/intechopen.74869

[4] Engeldrum PG. A theory of image quality: The image quality circle. Journal of Imaging Science and Technology. 2004;**48**:447-456

[5] Yoo JJ, Cui G, Luo MR. Image-quality modelling of a mobile display under various ambient illuminations. In: SID Symposium Digest of Technical Papers (SID '09). 2009. pp. 915-918

[6] ICDM. Information Display Measurements Standard (IDMS). Version 1.03. SID; 2012. 52 p

[7] IEC TS 62341-6-5. Organic light emitting diode (OLED) displays—Part 6–5: Measuring methods of dynamic range properties; 2019. pp. 9-14

[8] Recommendation ITU-R BT.709-6: Parameter Values for the HDTV Standards for Production and International Programme Exchange; 2015. 3 p

[9] Recommendation ITU-R BT.2020-2: Parameter Values for Ultra-High Definition Television Systems for Production and International Programme Exchange; 2015. 3 p

[10] SMPTE. Digital Source Processing—Color Processing for D-Cinema. EG 432-1:2010—SMPTE Engineering Guideline. pp. 431-432. DOI: 10.5594/SMPTE.EG432-1.2010

[11] Ye Z, Qiu J, Xu H, Luo MR, Westland S. Image quality evaluation of HDR displays. In: SID Symposium Digest of Technical Papers (SID '17). 2017. pp. 1192-1195

[12] Fairchild MD. Color Appearance Models. 3rd ed. Chichester: Wiley; 2013. pp. 119-120. DOI: 10.1002/9781118653128

[13] Reinhard E, Ward G, Debevec P, Pattanaik S. High Dynamic Range Imaging: Acquisition, Display, and Image-Based Lighting. 2nd ed. Burlington: Elsevier; 2010. pp. 1-10

[14] Wikipedia. High Dynamic Range. 2019. Available from: https://en.wikipedia.org/wiki/High_dynamic_range [Accessed: 15 March 2018]

[15] Hood DC, Finkelstein MA. Visual sensitivity. In: Boff K, Kaufman L, Thomas J, editors. Handbook of Perception and Human Performance; Volume 1: Sensory Processes and Perception. New York: Wiley; 1986

[16] SMPTE. High Dynamic Range Electro-Optical Transfer Function of Mastering Reference Displays. SMPTE ST. 2084; 2014. pp. 5-6. DOI: 10.5594/SMPTE.ST2084.2014

[17] Fang J, Xu H, Lv W, Luo MR. Proper luminance of HDR TV system. In: SID Symposium Digest of Technical Papers (SID '16). 2016. pp. 806-808

[18] Kwon J, Bang S, Kang D, Yoo JJ. The required attribute of displays for high dynamic range. In: SID Symposium Digest of Technical Papers (SID '16). 2016. pp. 884-887

[19] Park Y, Kwon J. Consideration of display metrology for HDR and WCG

standards based on real content. In: SID Symposium Digest of Technical Papers (SID '17). 2017. pp. 923-926

[20] Hwang JS, Park SH, Yoo JJ. Sound quality evaluation of OLED panel speaker. In: International Meeting on Information Display (IMID '18). 2017. 103 p

[21] EBU. Assessment Methods for the Quality of Sound Material—Music. EBU TECH 3286; 1997

[22] Sean O, Welti T, McMullin E. Listener preferences for in-room loudspeaker and headphone target responses. AES Convention. 2013;**135**: 8994

Chapter 6

TADF Technology for Efficient Blue OLEDs: Status and Challenges from an Industrial Point of View

Alhama Arjona-Esteban, Barbara Szafranowska and Julian Ochsmann

Abstract

The rise of OLED technology for display applications over the past decade was impressive. Today, OLED displays can be found everywhere, for example, in smartphones, TVs, smartwatches, monitors, cars, or digital cameras. However, as technology advances, the need for better OLED materials which help to improve energy efficiency and resolution of OLED displays is growing. While for the red and green pixels, phosphorescent materials have allowed for a boost in performance, the use of fluorescent materials for the blue pixel still limits the efficiency of OLED displays. Academic research has demonstrated many improvements regarding the efficiency of blue OLEDs using phosphorescent or TADF materials. However, studies on the limitations of device lifetime are rare. In the present chapter, the development of blue OLEDs based on TADF emitters is discussed from an industrial point of view. First, the material design principles for TADF molecules as well as the requirements for efficient blue TADF emitters are discussed. Moreover, a short literature overview on the latest improvements in blue TADF materials in academia and industry is presented. Finally, an outlook on this technology, its industrial possibilities, and alternatives is given.

Keywords: TADF, triplet up-conversion, industry, deep blue, high efficiency blue

1. Introduction

After their first development in the late 1980s [1], organic-light-emitting diodes slowly made their entry into the market of displays for televisions and mobile phones. Compared to other display technologies, OLED displays offer advantages such as wide viewing angles and very high contrast ratio while maintaining low energy consumption. In addition, OLED displays are thin and lightweight and can even be fabricated on flexible substrates, which allows for new and exciting applications. OLEDs can be produced using small molecules or polymers by vacuum thermal evaporation or by inkjet printing.

The basic pixels used to generate the primary colors, red, green, and blue, in an OLED display should have similar performance in order to guarantee a balanced efficiency and power consumption, which is beneficial for display technology, for example, in terms of increasing resolution and decreasing backplane technology complexity. However, as of today, the red and green pixels outperform the blue ones

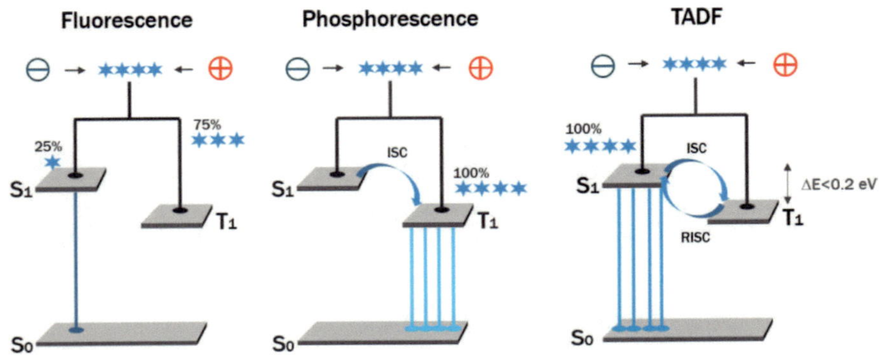

Figure 1.
Schematic energy diagrams of fluorescent, phosphorescent, and TADF molecules and their corresponding excitation and de-excitation pathways, when used in OLEDs. Fluorescence often relates to the first emitter material generation. Phosphorescence is the second material generation. TADF marks the latest conceptual development. Figure courtesy of CYNORA.

by far, which lies in the different technologies used for the different colors: There are three main photophysical mechanisms, which are relevant for light generation in OLEDs: fluorescence, phosphorescence, and thermally activated delayed fluorescence (TADF). Fluorescent materials were the first to be used in OLEDs [2]. However, the efficiency of a simple OLED device based on fluorescent materials is limited. These materials can only convert 25% of the excitons formed under electronic excitation into light because quantum mechanics dictates that under electronic excitation, singlet and triplet excitons are formed in a ratio of 1:3 (see **Figure 1**). Thus, with the discovery of phosphorescent materials, which can make use of all the excited states including triplets, the efficiency of green and red OLEDs increased drastically [3, 4]. However, as of today, no deep blue phosphorescent materials have been found to be stable enough for industrial applications. In this regard, the third generation of OLED materials comes into play. TADF technology allows, like phosphorescent materials, for the conversion of all excitons into light, but holds the promise of longer device lifetime [5–11].

In this chapter, we will focus on TADF technology as a potential alternative to fluorescent materials in deep blue OLEDs. We will discuss the advantages of this technology and the challenges connected to entering the industrial production of displays.

2. Requirements for TADF behavior and key performance indicators of TADF emitters

In this section, we will focus on the main design principles for TADF molecules and discuss the parameters that make TADF emission possible and how to tune them.

In simple terms, thermally activated delayed fluorescence, as depicted in **Figure 1**, is a fluorescence mechanism based on the repopulation of the singlet state by reverse intersystem crossing (RISC) from the triplet state, triggered by thermal energy. Thus, the theoretical internal quantum yield (IQE) of an OLED based on TADF is 100%, comparable to that of phosphorescent emitters and much higher than the 25% achieved by fluorescent materials, neglecting any additional contributions from triplet-triplet annihilation (TTA).

For the TADF process to be possible, the energy difference between the triplet and the singlet states (ΔE_{ST}) has to be smaller than 0.2 eV [12]. From the molecular design point of view, this small ΔE_{ST} is mostly addressed by spatially separating the highest occupied molecular orbital (HOMO) and lowest unoccupied molecular orbital (LUMO) of the molecule. The use of electron-donating (donor) and electron-withdrawing (acceptor) building blocks generates a charge-transfer structure with localized HOMO and LUMO orbitals. This localization can be favored by increasing the twist angle between the donor and acceptor moiety. However, a minimal overlap of the orbitals is necessary for RISC to take place, so a trade-off between small ΔE_{ST} and overlap has to be found.

Besides having a small ΔE_{ST}, an efficient TADF emitter has to fulfill other requirements: High photoluminescence quantum yields (PLQY) together with short excited state lifetimes ensure efficient RISC and radiative decay from the singlet state. Furthermore, a narrow spectrum at the desired color coordinate is necessary to provide the requested color purity. In addition, and especially for industrial applications, a long-lived device has to be delivered in order to satisfy the customers.

2.1 Color tuning: how to achieve blue

The use of strong donor and acceptor units is very widespread when aiming for small ΔE_{ST} and short excited state lifetime. However, in order to achieve blue emission, the electron-donor and acceptor units have to be carefully matched.

Since the frontier molecular orbitals (FMOs) in TADF emitters are highly localized on the donor and acceptor moieties, it can be assumed that the FMOs of the emitters will be very similar to the FMOs of the isolated donor and acceptor building blocks. Hence, the HOMO of the TADF emitter is determined by the donor moiety and the LUMO by the acceptor moiety (see **Figure 2**).

We can get an estimate on the donor and acceptor strength of different building blocks by performing density functional theory (DFT) calculations and comparing their HOMO and LUMO values. A higher HOMO level (less negative) corresponds to a stronger donor, while a strong acceptor has a lower LUMO level

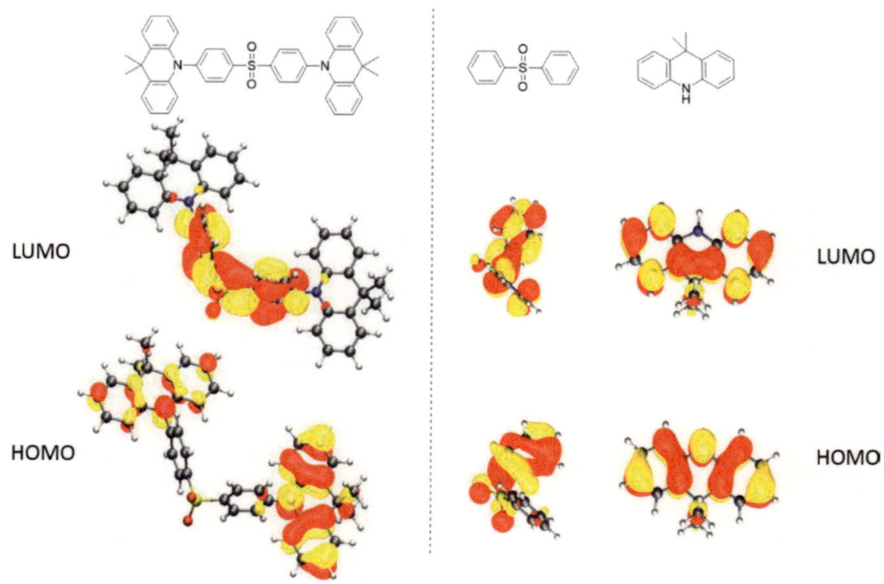

Figure 2.
The calculated FMO levels of the TADF emitter depicted on the left can be attributed to the respective HOMO (donor) and LUMO (acceptor) of the building blocks depicted on the right.

(more negative) (see **Figure 3**). A detailed list of the calculated HOMO and LUMO levels of many donor and acceptor building blocks has been published elsewhere [10].

Finally, from the difference between the calculated HOMO and LUMO levels, we can get a rough estimate of the optical bandgap, which corresponds to the combination of these donor and acceptor moieties. Carefully choosing the donor and acceptor units based on the calculations will allow us to reach the desired color range, in this case, deep blue with peak emission between 450 and 470 nm.

2.2 Narrow emission

Typically, TADF emitters display broad emission spectra. The reason is that the intramolecular charge-transfer (CT) state, where the emission originates, can undergo geometric relaxation, leading to many molecular configurations emitting at different energies. While for lighting applications, this intrinsic property of TADF emitters can be seen as an advantage, for display applications, the broad emission should be avoided. The display industry requires high color purity with CIEy values around 0.1 for the blue pixel, which can, with current host materials, only be achieved with a narrow spectrum. Moreover, when the spectrum is narrow, the emission onset can be shifted to lower energies, improving the blue color. Shifting the emission onset to lower energies is beneficial in matching the emitter with available host materials and getting a higher stability of the whole system (see Section 4: Requirements for host materials).

The host material can also influence the width and position of the spectrum of the TADF emitter. Polar hosts, for instance, often lead to broader and red-shifted spectra [13]. However, this effect is small in comparison to the intrinsic variations of the different emitter families [14]. Therefore, it is worth aiming to produce molecular structures that have an intrinsically smaller full width at half maximum (FWHM). In the literature, this is mainly addressed by rigidification of the donor-acceptor scaffold in order to minimize the rotational and bending modes and thereby suppressing the geometric relaxation of the CT state [10]. The rigidity can

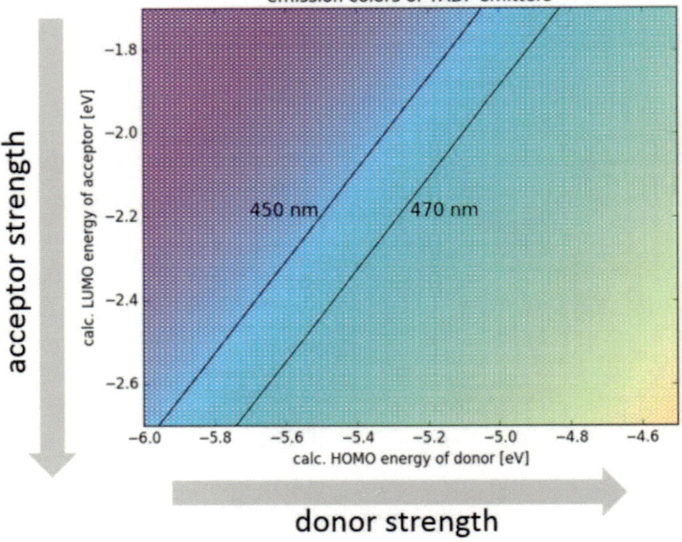

Figure 3.
Correlation of calculated HOMO and LUMO values of donor and acceptor building blocks with the expected emission color of the resulting TADF emitter. Figure courtesy of CYNORA.

be obtained by fusing the donor and acceptor units or introducing steric hindrances, as demonstrated in Ref. [15] by interlocking the donor units.

2.3 Tuning the excited state dynamics of TADF emitters

The main parameters which describe the excited state dynamics of TADF emitters are the fluorescence radiative decay rate k_F, the rate of delayed fluorescence k_{DF}, the rates of intersystem crossing k_{ISC}, and reverse intersystem crossing k_{RISC}, as well as the energy difference ΔE_{ST} between the first excited singlet S_1 and the first excited triplet T_1 states (see **Figure 4**). In order to obtain the best performing TADF emitters, all of these parameters have to be taken into account and careful molecular design must go hand in hand with in-depth photophysical analysis of these emitter systems.

The light-emitting performance of TADF materials depends on efficient radiative decay processes (k_F and k_{DF}) from S_1 to the electronic ground state S_0 and is also closely related to ΔE_{ST}, which governs the up-conversion of excitons from T_1 to S_1. A small ΔE_{ST} is crucial to obtain high values of k_{RISC}, improving the triplet harvesting through the thermally activated reverse intersystem crossing mechanism.

Decreasing the singlet-triplet energy gap ΔE_{ST} and increasing the radiative decay rate k_F is a useful way to get efficient TADF emission. Unfortunately, as mentioned at the beginning of Section 2, there is a trade-off between getting a small ΔE_{ST} and getting a high k_F, which is related to the electronic coupling between the ground state and the singlet excited state. A good separation of the HOMO and LUMO orbitals leads to a small ΔE_{ST}, but it also reduces k_F [16]. On the contrary, compounds with overlapping HOMO and LUMO orbitals are expected to exhibit high values for k_F, but yielding relatively large ΔE_{ST}, and thus showing less TADF contribution [17].

One of the approaches to increase the radiative decay rate, while maintaining a small ΔE_{ST} in TADF emitters, is the concept of using an aromatic bridge or spacer between the donor and acceptor moieties in the molecular design. In such a system, the delocalization of molecular orbitals is given, while there is in addition a weak overlap of HOMO and LUMO orbitals at the spacer, which induces a larger oscillator strength, enhancing the rate of radiative decay. This concept was successfully demonstrated in many TADF emitters containing phenyl linkers between donor and acceptor units in one TADF molecule [17–20]. Initially, the aromatic bridge negatively influenced ΔE_{ST}—making it larger—because of an increased S_1 energy level, whereas the T_1 level was reduced by the extended π-conjugation. But in such cases, the singlet-triplet energy gap could still be reduced by using stronger donors and acceptors and thereby strengthening the charge-transfer character [21–23]. In addition, the π-conjugation can be interrupted by twisting the donor and acceptor

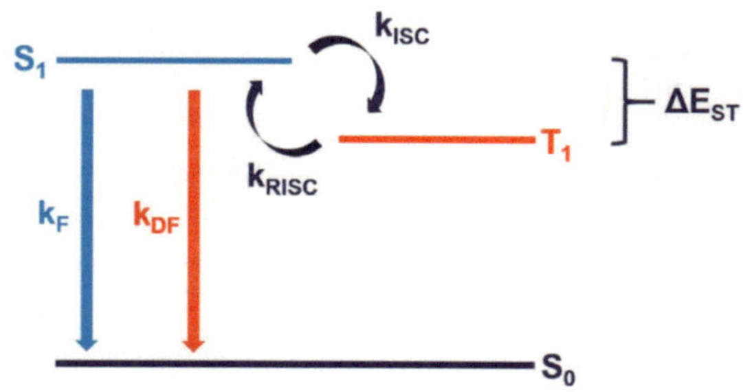

Figure 4.
Schematic energy diagram showing the main parameters which determine the excited state dynamics of TADF emitters.

units, so that they are nearly orthogonal to each other. This design strategy is also very beneficial in decreasing ΔE_{ST} and hence increasing k_{RISC}.

Other approaches reported in the literature to get highly efficient TADF materials are to delocalize the HOMO by increasing the number of donors [10, 18, 20] or to design a dual emitting core [24, 25]. In the second case, light absorption and emission is intensified by introducing multiple TADF chromophores fused into one molecule, yielding high PLQY.

The rate of reverse intersystem crossing k_{RISC} is directly linked to the easily observable excited state lifetime of delayed fluorescence τ_{DF}, which is usually in the order of a few microseconds for efficient TADF molecules. The higher the k_{RISC}, the shorter is τ_{DF}. τ_{DF} is a measure for how long on average the triplet excited state of the TADF emitter is populated. Long-lived triplet excitons are believed to be destructive for OLED devices, because they enable potentially detrimental processes such as triplet-triplet or triplet-polaron annihilation [26–29]. Therefore, triplet excited state lifetimes of TADF emitters should be as short as possible to enhance device stability by reducing detrimental quenching processes. Again, the reduction of the ΔE_{ST} gap results in speeding up the triplet harvesting mechanism. Several new molecular design methods have been proposed to reach an excited state lifetime of only a few microseconds, including physical separation of donor and acceptor units, dual emitting cores, breaking conjugation by twisted structures, multi-donor approach, introduction of strong donors and acceptors, and others described above [30, 31].

3. Blue TADF emitters in the literature

The general approach followed in the literature for the design of TADF emitters consists in combining donor and acceptor moieties into one molecule to separate the HOMO and LUMO orbitals and ensure a small ΔE_{ST} value. In this section, the latest developments reported in the literature since 2017 are collected and classified by

Compound	PI	Performance	Group
Sf-3Cz	Fan, Liao and coworkers [36]	15% EQE at 100 cd/m^2 CIE (0.16, 0.14)	Sulfone
G2	Xu and coworkers [35]	(no EQE) CIE (0.15, 0.12)	Sulfone
3DpYM-pDTC	Cheng and coworkers [37]	20% EQE at 500 cd/m^2 CIE (0.14, 0.18)	Ketone
TXAZ	Kim, Kim and coworkers [38]	16% EQE max (DPEPO) CIE (0.15, 0.13)	Triazine
FATA	Kaji and coworkers [34]	11% EQE max CIE (0.15, 0.13)	Triazine
Ac3MHPM	Kido and coworkers [39]	10% EQE at 100 cd/m^2 CIE (0.16, 0.15)	Pyrimidine
CNICtCz	Lee and coworkers [40]	10% EQE at 100 cd/m^2 CIE(0.14, 0.13)	Nitrile
DCzBN3	Adachi and coworkers [41]	10% EQE at 100 cd/m^2 CIE (0.15, 0.06)	Nitrile
TN4T-PCZ	Zheng, Meng and coworkers [33]	20% EQE max CIE (0.16, 0.03)	CF3-pyridine

Table 1.
Status of deep-blue TADF technology as published in the literature.

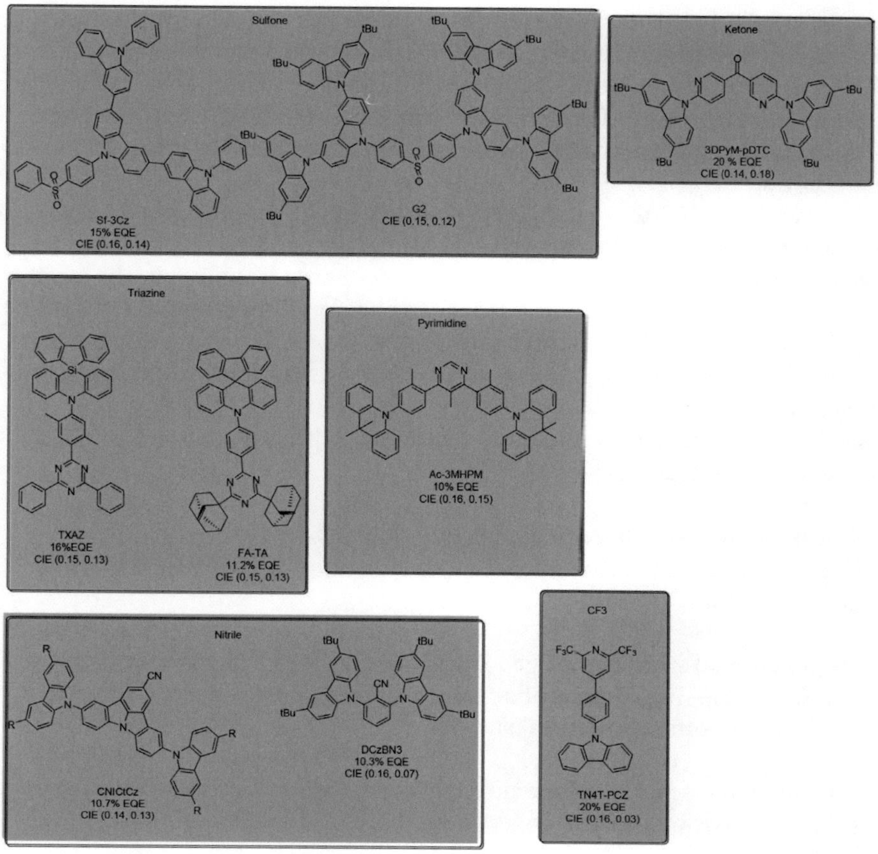

Figure 5.
Selected structures of published deep-blue TADF emitters.

the acceptor unit. We have focused on efficient devices (EQE > 10%) with deep-blue color (CIEy < 0.15). The references are collected in **Table 1** and **Figure 5**. For earlier results, we recommend other review articles [9–11, 32]. Most of the collected articles put their main focus on achieving deep blue color by combining donor units based on carbazole or acridine motifs and acceptors such as sulfones, triazines, or pyrimidines. While high EQE values can be achieved [33], these data have to be taken with care, since often they are given at low luminance (10–100 cd/m^2) and the devices suffer from high roll-off. Unfortunately, very few data are provided on device stability.

Interestingly, many approaches aim for solution-processed OLEDs either using small molecules bearing solubilizing chains [34] or dendrimer structures [35]. Currently, the vast majority of OLED products issued from the display industry are based on the evaporation of small molecules under vacuum. Printing and coating techniques are however receiving increasing attention due to the evident advantages of cost-effectivity and low material consumption.

4. Requirements for host materials

It is commonly known that the performance of TADF emitters is often very poor in neat emitter films, because of concentration quenching, aggregation, and other effects. Therefore, diluting the emitter molecules with host molecules at relatively low emitter concentrations (10–30%) improves not only their photophysical characteristics but also boosts the efficiency in devices [42]. One of the basic requirements for a host material is a sufficiently high triplet state energy. In fact, the T_1-state of the host is ideally higher than the energy of the T_1-state of the TADF emitter in order to prevent Dexter energy transfer of triplet excited states from the TADF emitter to the host material, which represents a major loss channel in TADF-based OLEDs.

Moreover, the HOMO- and LUMO-energy level differences between the host and emitter material need to be well adjusted. Large energy level differences between the host and emitter may cause direct charge trapping on the emitter or unwanted exciplex formation. The energy levels of the materials have to be carefully aligned in order to achieve the desired energy and/or charge carrier transfer between those two species, which helps in improving light-emitting performance.

Besides these two important points described above, a host for TADF should also have the following features [28, 43, 44]:

- **Thermal and morphological stability**: defined by the high glass transition temperature (T_g) for stabilizing the film morphology and inhibiting crystallization.

- **High material stability**: judged mainly by the bond dissociation energies (BDEs) of the single bonds of a molecule under three different conditions (cationic, anionic and neutral state).

- **Low polarity**: to reduce the emitter-host local dipole interactions that tend to shift the emission peak to longer wavelengths (bathochromic shift).

Furthermore, the triplet energy T_1 is sensitive towards the following [43, 45–48]:

- **Extended π-conjugation**: triplet energy decreases considerably in the line of benzene, biphenyl, and p-terphenyl. Also, fused ring systems lower the triplet energy (naphthalene less than biphenyl).

- **Connection pattern and torsion angles**: highly twisted molecular structures reduce the conjugation, increasing the triplet level; *meta*-linkage leads to a lower degree of conjugation and is more beneficial to get high triplet levels than *para*-linkage.

- **Aggregation**: bulky substituents weaken intermolecular reactions of π-conjugated host molecules, thus reducing the red-shift of phosphorescence spectra (T_1 increase). Hence, depending on the strength of interactions, T_1 measured in solution can be higher compared to those from neat films [49].

Stable hosts for blue TADF emitters are still very rare, because they require triplet energies above 3.0 eV to guarantee an efficient exciton confinement. Versatile building blocks like carbazole, thiophene, or benzofuran are often used for host and emitter materials targeted for deep blue, because of their high intrinsic triplet energies of 2.8–3.0 eV.

One of the most popular hosts for blue TADF is DPEPO, benefiting from the high triplet energy of 3.00 eV [50]. Unfortunately, OLED devices with DPEPO generally suffer from serious efficiency roll-off due to its poor stability [51]. Other conventional host examples classified by their potential charge transport characteristics are shown below [43, 52, 53] (**Figure 6**):

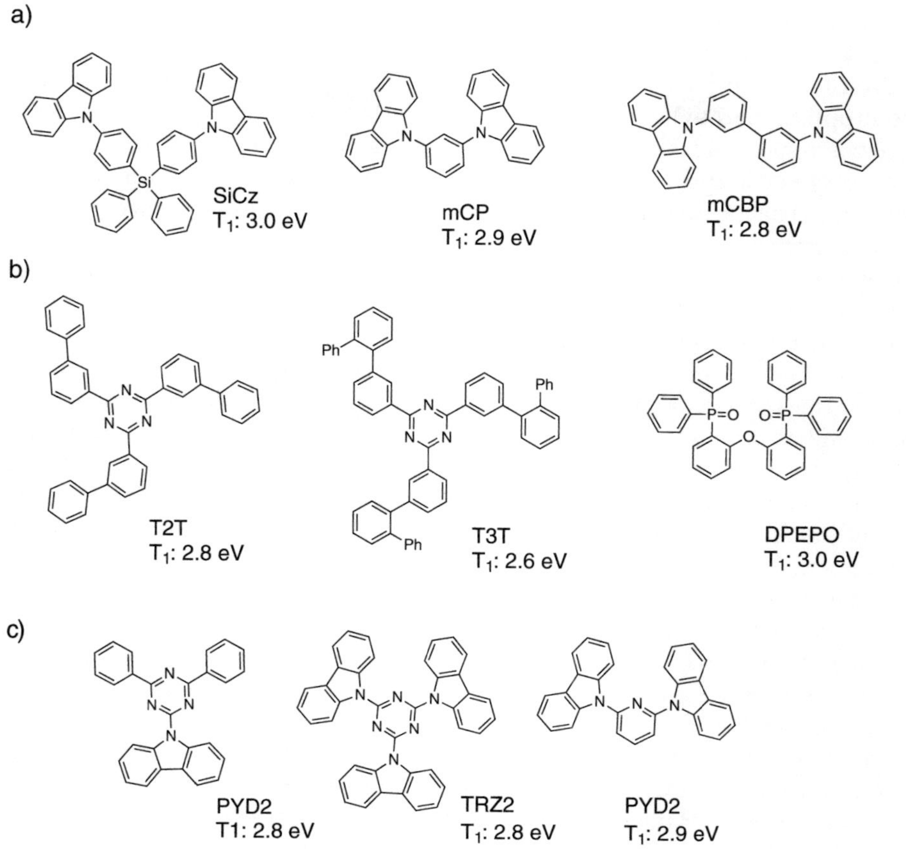

Figure 6.
Different host materials reported in the literature classified by their potential charge transport characteristics. (a) p-type hosts (hole transport character). (b) n-type hosts (electron transport character). (c) Ambipolar hosts (hole and electron transport character).

5. Next generation OLED: hyper-fluorescence

In the last few years, TADF emitters have become very popular for application in OLEDs, because of their triplet up-conversion property, which allows for high external quantum efficiency. In fact, TADF emitters can deliver device efficiencies, which can be up to three times higher than for OLEDs based on fluorescent dopants. However, many highly efficient TADF materials suffer from relatively strong degradation (e.g., due to long exciton lifetimes) and from broad emission spectra (characteristic for the intramolecular charge transfer states). One recently reported strategy to overcome those hurdles is the **hyper-fluorescence** approach (which is sometimes also referred to as TADF assisted fluorescence), first demonstrated by Adachi in 2014 [54].

In this approach, the outstanding properties of TADF molecules are utilized in combination with the narrow emission of fluorescent molecules: the TADF emitter serves as the triplet up-conversion material, which is responsible for high efficiency. After the up-conversion of triplet excited states to the singlet state of the TADF molecule, these singlet excited states can be transferred via Förster resonance energy transfer (FRET) to the S_1 state of the fluorescent emitter. Ideally, light emission occurs exclusively from the fluorescent dopant, which delivers a well-defined color point (**Figure 7**). If both, the rates of RISC and FRET are high, the overall excited state lifetime in this system is decreased, and hence the operational lifetime of the whole system is expected to be increased.

Hyper-fluorescence with all of its potential benefits is a very promising technology, which is not easy to engineer, because many criteria have to be met [54–58]. The two main criteria, which have to be fulfilled in order to achieve efficient hyper-fluorescence are an optimal spectral overlap of the TADF molecule's emission spectrum with the absorption spectrum of the fluorescent molecule and a high PLQY of both molecules. Achieving a great spectral overlap while maintaining a high device efficiency is especially challenging for hyper-fluorescence in the deep blue color range, where fluorescent materials with absorption peaks in the range of 400–450 nm and small Stokes shifts have to be combined with blue TADF emitters [57]. In addition, the TADF component should exhibit very fast reverse intersystem crossing from the triplet to the singlet excited state [57].

Moreover, the unwanted short-range Dexter energy transfer from the triplet state of the TADF emitter to the non-emitting triplet state of the fluorescent emitter should be avoided. Since Dexter transfer is highly dependent on the distance between the involved species, one way to minimize it is to physically separate the fluorescent and TADF emitting cores. This can be done by reducing the concentration of the fluorescent material in the system (typically ~1 wt.%) or by shielding the fluorescent dopant via the introduction of bulky side groups [55, 59]. This kind of molecular shielding is also beneficial to prevent self-quenching when using higher doping concentrations of the fluorescent material [60].

Finally, it is also important to align the HOMO and LUMO levels of all components within the emission layer (EML) in order to reduce charge trapping by the fluorescent material.

Besides using a conventional fluorescent molecule in such a "hyper-approach", where two emitters are combined, one can also use DABNA molecules, which were first introduced by Hatakeyama in 2016 [61]. DABNA derivatives are very interesting emitters for display applications in a hyper-approach, since they have even narrower emission and smaller Stokes shifts than most conventional fluorescent dopants.

Summarizing, hyper-fluorescence is a very complex approach where three or four materials are mixed within the EML and whose mechanisms are not yet fully understood. However, and despite the difficult and limited material selection, especially for the deep blue region, this technology is considered a promising

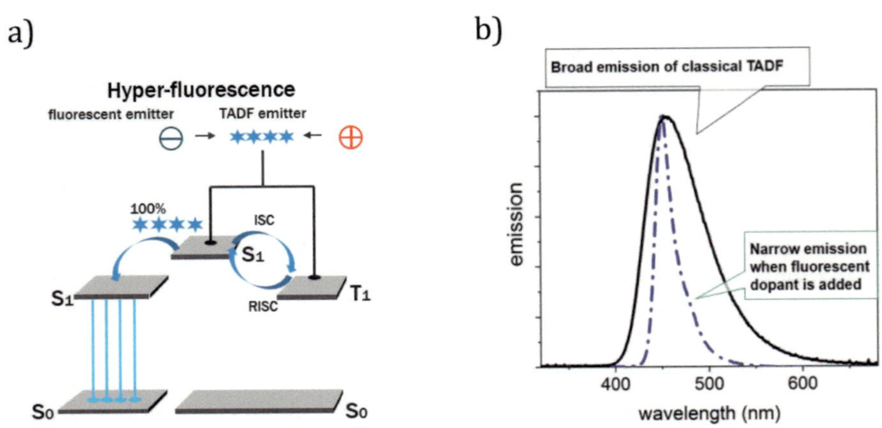

Figure 7.
(a) A simplified hyper-fluorescent diagram of a three-component system. The excitons are generated on the TADF material. The triplets are up-converted to singlets, which are further transferred via Förster resonance energy transfer (FRET) to the fluorescent dopant. The radiative decay to the ground state takes place on the fluorescent dopant. The host is omitted for clarity. (b) The FWHM of the emitting system based on TADF is significantly reduced when doped with a fluorescent material. Figure courtesy of CYNORA.

alternative for TADF and phosphorescence technologies, and holds the promise of high efficiency and stability as required by the display industry.

6. Industry status: Kyulux and CYNORA

Table 2 summarizes some of the device performances, which were published by CYNORA and one result announced by Kyulux. While the efficiency is usually reported at a similar brightness (1000 cd/m^2), comparing the results for color and device lifetime is not as easy, since color cannot be accurately defined by peak emission wavelength alone and device lifetime is often measured under different conditions depending on the preference of the company. A peak emission wavelength of ~460 nm is required for deep-blue OLED pixels, but this requirement is not sufficient to define the color. Additional parameters that also include the shape of the emission spectrum are the CIE 1931 coordinates CIEx and CIEy, which define a point in color space. For the deep blue region, it is usually sufficient to only look at the CIEy coordinate, which needs to be ≤0.15.

Device lifetimes are determined by measuring the decrease of luminance over time until a certain loss of the initial luminance is reached. If the luminance decay of a device was measured starting at a luminance of 750 cd/m^2 until the time were the luminance decreased by 5%, this lifetime value would be given as LT95 at 750 cd/m^2. Lifetimes measured under different conditions can be converted into other lifetime values. However, there is always an error connected to such a calculation. In general, it will always be the case that the lifetime is reduced for increasing brightness, and the lifetime is increased when a larger decrease in luminance is measured (LT90 is always longer than LT95).

In May 2017, CYNORA announced the result of a sky blue OLED with 15% EQE, a CIEy of 0.28, a peak emission of around 470 nm, and a LT95 at 750 cd/m^2 of around 190 hours. This was an important milestone on the way to an efficient deep blue TADF emitter. Before this result, TADF lifetimes LT95 for sky blue and deep blue had usually been in the minutes.

In December of the same year, CYNORA could build highly efficient OLED devices having 24% EQE with a deep blue TADF emitter at a CIEy of 0.15 and a peak emission of 460 nm. However, the device lifetime in this system was also reduced by roughly one order of magnitude to 30 hours. These two results impressively show the difficulty of reaching deep blue emission color together with high efficiency, and maintaining or improving device lifetime at the same time. These three factors,

	EQE at 1000 cd/m^2 (%)	CIEy	Peak emission (nm)	LT95 at 750 cd/m^2 (h)	Date
Deep-blue color requirement		≤0.15	~460		
CYNORA	15	0.28	<470	~190	May, 2017
	24	0.15	460	~30	Dec, 2017
	20	0.13	464	~15	August, 2018
Kyulux	22		470	~100	July, 2018 [62]

Table 2.
Status of Kyulux and CYNORA.

efficiency, color and lifetime, are clearly very closely linked, which means that device performance can only be compared between devices with the same color. Later in 2018, CYNORA also introduced results based on the hyper-approach. Even a lower CIEy of .013 at an emission wavelength of 464 nm together with a high EQE of 20% and a - for this color point- decent device lifetime of 15 hours LT95 at 750 cd/m^2 were achieved.

Also, in 2018 (July), Kyulux announced a first sky blue OLED device result based on a hyper-approach with an emission maximum at 470 nm, an EQE of 22% and an LT95 at 750 cd/m^2 of 100 hours.

Clearly, the results in the field of TADF of both companies, Kyulux and CYNORA, are impressive. Within just a few years, the companies showed that the TADF technology could be advanced from an interesting R&D topic to a serious contender for the next generation of high-efficiency emitters. The challenge for CYNORA is now to improve the stability of the TADF-based blue OLED pixel to a level that makes mass production with this technology feasible. A highly efficient and stable blue emitter will solve many technical issues for the display makers and pave the way for displays with lower power consumption and for higher display resolution.

7. Conclusion: what needs to be solved to get TADF ready?

From the latest improvements in academia and industry, it is clear that TADF technology is capable of achieving the desired blue color point necessary to succeed in the display field. Many TADF-based systems with deep blue color and high efficiency have been reported [9–11, 32]. However, challenges when going to deeper blue color remain the availability of suitable host materials with sufficiently high triplet levels and device stability. A way to circumvent the issue in the deep blue color region of having an emission onset at low wavelengths, that is, high energies, is to combine TADF emitters with narrow-emitting fluorescent dopants in the hyper-approach, where the excitons formed on the TADF material transfer their energy via FRET to the fluorescent emitter. Again, while this approach holds the promise of very narrow and efficient emission, engineering a three-component emissive layer requires the matching of several materials in terms of absorption and emission spectra as well as frontier molecular orbital alignment.

Finally, the most pressing issue and the one receiving the least attention in the literature remains the intrinsic stability of the materials, including the host, which is the material used at the highest concentration in the emissive layer, the TADF emitter, which determines the final efficiency of the device, and all stack materials which can potentially affect the device stability. From a molecular design point of view, chemically stable bonds are preferred. Moreover, in the device stack, the energy levels of all compounds need to be optimized in order to prevent too much stress on a particular material, which might lead to decomposition. This challenge is the focus of research-driven companies like CYNORA. After the fast development of the last years, we are confident that this advancement is achievable in a short period of time and that TADF technology will be ready for application soon.

Author details

Alhama Arjona-Esteban*, Barbara Szafranowska and Julian Ochsmann
cynora GmbH, Bruchsal, Germany

*Address all correspondence to: esteban@cynora.com

References

[1] Tang CW, Van Slyke SA. Organic electroluminescent diodes. Applied Physics Letters. 1987;**51**:913. DOI: 10.1063/1.98799

[2] Van Slyke SA, Chen CH, Tang CW. Organic electroluminescent devices with improved stability. Applied Physics Letters. 1996;**69**:2160. DOI: 10.1063/1.117151

[3] Baldo MA, O'Brien DF, You Y, Shoustikov A, Sibley S, Thompson ME, et al. Highly efficient phosphorescent emission from organic electroluminescent devices. Nature. 1998;**395**:151-154. DOI: 10.1038/25954

[4] Baldo MA, Lamansky S, Burrows PE, Thompson ME, Forrest SR. Very high-efficiency green organic light-emitting devices based on electrophosphorescence. Applied Physics Letters. 1999;**75**:4-6. DOI: 10.1063/1.124258

[5] Endo A, Sato K, Yoshimura K, Kai T, Kawada A, Miyazaki H, et al. Efficient up-conversion of triplet excitons into a singlet state and its application for organic light emitting diodes. Applied Physics Letters. 2011;**98**:083302. DOI: 10.1063/1.3558906

[6] Bergmann L, Zink DM, Bräse S, Baumann T, Volz D. Metal-organic and organic TADF-materials: Status, challenges and characterization. Topics in Current Chemistry. 2016;**374**:22. DOI: 10.1007/s41061-016-0022-6

[7] Adachi C. Third-generation organic electroluminescence materials. Japanese Journal of Applied Physics. 2014;**53**:060101. DOI: 10.7567/JJAP.53.060101

[8] Volz D. Review of organic light-emitting diodes with thermally activated delayed fluorescence emitters for energy-efficient sustainable light sources and displays. Journal of Photonics for Energy. 2016;**6**:020901. DOI: 10.1117/1.JPE.6.020901

[9] Cai X, Su S-JJ. Marching toward highly efficient, pure-blue, and stable thermally activated delayed fluorescent organic light-emitting diodes. Advanced Functional Materials. 2018;**1802558**: 1-33. DOI: 10.1002/adfm.201802558

[10] Im Y, Kim M, Cho YJ, Seo JA, Yook KS, Lee JY. Molecular design strategy of organic thermally activated delayed fluorescence emitters. Chemistry of Materials. 2017;**29**:1946-1963. DOI: 10.1021/acs.chemmater.6b05324

[11] Bui T-T, Goubard F, Ibrahim-Ouali M, Gigmes D, Dumur F. Thermally activated delayed fluorescence emitters for deep blue organic light emitting diodes: A review of recent advances. Applied Sciences. 2018;**8**:494. DOI: 10.3390/app8040494

[12] Dias FB. Kinetics of thermal-assisted delayed fluorescence in blue organic emitters with large singlet-triplet energy gap. Philosophical Transactions of the Royal Society A: Mathematical Physical and Engineering Sciences. 2015;**373**:20140447-20140447. DOI: 10.1098/rsta.2014.0447

[13] Santos PL, Ward JS, Data P, Batsanov AS, Bryce MR, Dias FB, et al. Engineering the singlet-triplet energy splitting in a TADF molecule. Journal of Materials Chemistry C. 2016;**4**:3815-3824. DOI: 10.1039/C5TC03849A

[14] Noriega R, Barnard ES, Ursprung B, Cotts BL, Penwell SB, Schuck PJ, et al. Uncovering single-molecule photophysical heterogeneity of bright, thermally activated delayed fluorescence emitters dispersed in glassy hosts. Journal of the American Chemical

Society. 2016;**138**:13551-13560. DOI: 10.1021/jacs.6b05488

[15] Cho YJ, Jeon SK, Lee SS, Yu E, Lee JY. Donor interlocked molecular design for fluorescence-like narrow emission in deep blue thermally activated delayed fluorescent emitters. Chemistry of Materials. 2016;**28**:5400-5405. DOI: 10.1021/acs.chemmater.6b01484

[16] Berberan-Santos MN, Garcia JMM. Unusually strong delayed fluorescence of C_{70}. Journal of the American Chemical Society. 1996;**118**:9391-9394. DOI: 10.1021/ja961782s

[17] Hirata S, Sakai Y, Masui K, Tanaka H, Lee SY, Nomura H, et al. Highly efficient blue electroluminescence based on thermally activated delayed fluorescence. Nature Materials. 2015;**14**:330-336. DOI: 10.1038/nmat4154

[18] Uoyama H, Goushi K, Shizu K, Nomura H, Adachi C. Highly efficient organic light-emitting diodes from delayed fluorescence. Nature. 2012;**492**:234-238. DOI: 10.1038/nature11687

[19] Lee DR, Kim BS, Lee CW, Im Y, Yook KS, Hwang SH, et al. Above 30% external quantum efficiency in green delayed fluorescent organic light-emitting diodes. ACS Applied Materials & Interfaces. 2015;**7**:9625-9629. DOI: 10.1021/acsami.5b01220

[20] Lee DR, Kim M, Jeon SK, Hwang SH, Lee CW, Lee JY. Design strategy for 25% external quantum efficiency in green and blue thermally activated delayed fluorescent devices. Advanced Materials. 2015;**27**:5861-5867. DOI: 10.1002/adma.201502053

[21] Zhang Q, Li B, Huang S, Nomura H, Tanaka H, Adachi C. Efficient blue organic light-emitting diodes employing thermally activated delayed fluorescence. Nature Photonics. 2014;**8**:326-332. DOI: 10.1038/nphoton.2014.12

[22] Baek JY, Lee J-H, Kwon S-K, Moon C-K, Kim K-H, Sun JW, et al. Thermally activated delayed fluorescence from azasiline based intramolecular charge-transfer emitter (DTPDDA) and a highly efficient blue light emitting diode. Chemistry of Materials. 2015;**27**:6675-6681. DOI: 10.1021/acs.chemmater.5b02515

[23] Shizu K, Tanaka H, Uejima M, Sato T, Tanaka K, Kaji H, et al. Strategy for designing electron donors for thermally activated delayed fluorescence emitters. Journal of Physical Chemistry C. 2015;**119**:1291-1297. DOI: 10.1021/jp511061t

[24] Cho YJ, Jeon SK, Chin BD, Yu E, Lee JY. The design of dual emitting cores for green thermally activated delayed fluorescent materials. Angewandte Chemie. 2015;**127**:5290-5293. DOI: 10.1002/ange.201412107

[25] Kim M, Jeon SK, Hwang S, Lee S, Yu E, Lee JY. Highly efficient and color tunable thermally activated delayed fluorescent emitters using a "twin emitter" molecular design. Chemical Communications. 2015;**52**:339-342. DOI: 10.1039/C5CC07999C

[26] Giebink NC, D'Andrade BW, Weaver MS, Brown JJ, Forrest SR. Direct evidence for degradation of polaron excited states in organic light emitting diodes. Journal of Applied Physics. 2009;**105**:124514. DOI: 10.1063/1.3151689

[27] Sandanayaka ASD, Matsushima T, Adachi C. Degradation mechanisms of organic light-emitting diodes based on thermally activated delayed fluorescence molecules. Journal of Physical Chemistry C.

2015;**119**:23845-23851. DOI: 10.1021/acs.jpcc.5b07084

[28] Song W, Lee JY. Degradation mechanism and lifetime improvement strategy for blue phosphorescent organic light-emitting diodes. Advanced Optical Materials. 2017;**5**:1600901. DOI: 10.1002/adom.201600901

[29] Wang Q, Aziz H. Degradation of organic/organic interfaces in organic light-emitting devices due to polaron-exciton interactions. ACS Applied Materials & Interfaces. 2013;**5**:8733-8739. DOI: 10.1021/am402537j

[30] Lee J, Shizu K, Tanaka H, Nakanotani H, Yasuda T, Kaji H, et al. Controlled emission colors and singlet-triplet energy gaps of dihydrophenazine-based thermally activated delayed fluorescence emitters. Journal of Materials Chemistry C. 2015;**3**:2175-2181. DOI: 10.1039/c4tc02530j

[31] Yang Z, Mao Z, Xie Z, Zhang Y, Liu S, Zhao J, et al. Recent advances in organic thermally activated delayed fluorescence materials. Chemical Society Reviews. 2017;**46**:915-1016. DOI: 10.1039/C6CS00368K

[32] Arjona-Esteban A, Volz D. Status and next steps of TADF technology: An industrial perspective. In: Highly Efficient OLEDs. Weinheim, Germany: Wiley-VCH Verlag GmbH & Co. KGaA; 2018. pp. 543-572. DOI: 10.1002/9783527691722.ch15

[33] Liang X, Han HB, Yan ZP, Liu L, Zheng YX, Meng H, et al. Versatile functionalization of trifluoromethyl based deep blue thermally activated delayed fluorescence materials for organic light emitting diodes. New Journal of Chemistry. 2018;**42**:4317-4323. DOI: 10.1039/c7nj04482h

[34] Wada Y, Kubo S, Kaji H. Adamantyl substitution strategy for realizing solution-processable thermally stable deep-blue thermally activated delayed fluorescence materials. Advanced Materials. 2018;**30**:1705641. DOI: 10.1002/adma.201705641

[35] Li J, Liao X, Xu H, Li L, Zhang J, Wang H, et al. Deep-blue thermally activated delayed fluorescence dendrimers with reduced singlet-triplet energy gap for low roll-off non-doped solution-processed organic light-emitting diodes. Dyes and Pigments. 2017;**140**:79-86. DOI: 10.1016/j.dyepig.2017.01.036

[36] Zhang W, Zhang Y-X, Zhang X-Q, Liu X-Y, Fan J, Liao L-S. Blue thermally activated delayed fluorescence materials based on bi/tri-carbazole derivatives. Organic Electronics. 2018;**58**:238-244. DOI: 10.1016/j.orgel.2018.04.022

[37] Rajamalli P, Senthilkumar N, Huang P-Y, Ren-Wu C-C, Lin H-W, Cheng C-H. New molecular design concurrently providing superior pure blue, thermally activated delayed fluorescence and optical out-coupling efficiencies. Journal of the American Chemical Society. 2017;**139**:10948-10951. DOI: 10.1021/jacs.7b03848

[38] Woo S-J, Kim Y, Kwon S-K, Kim Y-H, Kim J-J. Phenazasiline/spiroacridine donor combined with methyl-substituted linkers for efficient deep blue thermally activated delayed fluorescence emitters. ACS Applied Materials & Interfaces. 2019;**11**:7199-7207. DOI: 10.1021/acsami.8b20009

[39] Komatsu R, Ohsawa T, Sasabe H, Nakao K, Hayasaka Y, Kido J. Manipulating the electronic excited state energies of pyrimidine-based thermally activated delayed fluorescence emitters to realize efficient deep-blue emission. ACS Applied Materials & Interfaces. 2017;**9**:4742-4749. DOI: 10.1021/acsami.6b13482

[40] Im Y, Han SH, Lee JY. Deep blue thermally activated delayed

fluorescent emitters using CN-modified indolocarbazole as an acceptor and carbazole-derived donors. Journal of Materials Chemistry C. 2018;**6**:5012-5017. DOI: 10.1039/c8tc00546j

[41] Chan CY, Cui LS, Kim JU, Nakanotani H, Adachi C. Rational molecular design for deep-blue thermally activated delayed fluorescence emitters. Advanced Functional Materials. 2018;**28**:1-7. DOI: 10.1002/adfm.201706023

[42] Méhes G, Goushi K, Potscavage WJ, Adachi C. Influence of host matrix on thermally-activated delayed fluorescence: Effects on emission lifetime, photoluminescence quantum yield, and device performance. Organic Electronics. 2014;**15**:2027-2037. DOI: 10.1016/j.orgel.2014.05.027. https://www.sciencedirect.com/science/article/pii/S1566119914002067?via%3Dihub

[43] Tao Y, Yang C, Qin J. Organic host materials for phosphorescent organic light-emitting diodes. Chemical Society Reviews. 2011;**40**:2943-2970. DOI: 10.1039/c0cs00160k

[44] Zhang D, Cai M, Bin Z, Zhang Y, Zhang D, Duan L. Highly efficient blue thermally activated delayed fluorescent OLEDs with record-low driving voltages utilizing high triplet energy hosts with small singlet-triplet splittings. Chemical Science. 2016;**7**:3355-3363. DOI: 10.1039/C5SC04755B

[45] Bagnich SA, Rudnick A, Schroegel P, Strohriegl P, Köhler A. Triplet energies and excimer formation in *meta*- and *para*-linked carbazolebiphenyl matrix materials. Philosophical Transactions of the Royal Society A: Mathematical Physical and Engineering Sciences. 2015;**373**:20140446. DOI: 10.1098/rsta.2014.0446

[46] Strohriegl P, Wagner D, Schrögel P, Hoffmann ST, Köhler A, Heinemeyer U, et al. Novel host materials for blue phosphorescent OLEDs. In: So F, Adachi C, editors. 2013. p. 882906. DOI: 10.1117/12.2023305

[47] Oshiyama T, Yasukawa N, Kita H, Matsui T, Sumita M, Morihashi K. A theoretical guideline for designing effective host materials based on 4,4′-bis(9-carbazolyl)-1,1′-biphenyl derivatives for blue phosphorescent devices. Bulletin of the Chemical Society of Japan. 2017;**90**:195-204. DOI: 10.1246/bcsj.20160345

[48] Varathan E, Vijay D, Kumar PSV, Subramanian V. Computational design of high triplet energy host materials for phosphorescent blue emitters. Journal of Materials Chemistry C. 2013;**1**:4261-4274. DOI: 10.1039/c3tc30473f

[49] Woon KL, Hasan ZA, Ong BK, Ariffin A, Griniene R, Grigalevicius S, et al. Triplet states and energy back transfer of carbazole derivatives. RSC Advances. 2015;**5**:59960-59969. DOI: 10.1039/C5RA09340F

[50] Han C, Zhao Y, Xu H, Chen J, Deng Z, Ma D, et al. A simple phosphine-oxide host with a multi-insulating structure: High triplet energy level for efficient blue electrophosphorescence. Chemistry: A European Journal. 2011;**17**:5800-5803. DOI: 10.1002/chem.201100254

[51] Ihn S-G, Lee N, Jeon SO, Sim M, Kang H, Jung Y, et al. An alternative host material for long-lifespan blue organic light-emitting diodes using thermally activated delayed fluorescence. Advanced Science. 2017;**4**:1600502. DOI: 10.1002/advs.201600502

[52] Wong MY, Zysman-Colman E. Purely organic thermally activated delayed fluorescence materials for organic light-emitting diodes. Advanced Materials. 2017;**29**:1605444. DOI: 10.1002/adma.201605444

[53] Chen D, Su SJ, Cao Y. Nitrogen heterocycle-containing materials for highly efficient phosphorescent OLEDs with low operating voltage. Journal of Materials Chemistry C. 2014;**2**:9565-9578. DOI: 10.1039/c4tc01941e

[54] Nakanotani H, Higuchi T, Furukawa T, Masui K, Morimoto K, Numata M, et al. High-efficiency organic light-emitting diodes with fluorescent emitters. Nature Communications. 2014;**5**:4016. DOI: 10.1038/ncomms5016

[55] Zhang D, Song X, Cai M, Duan L. Blocking energy-loss pathways for ideal fluorescent organic light-emitting diodes with thermally activated delayed fluorescent sensitizers. Advanced Materials. 2018;**30**:1-10. DOI: 10.1002/adma.201705250

[56] Furukawa T, Nakanotani H, Inoue M, Adachi C. Dual enhancement of electroluminescence efficiency and operational stability by rapid upconversion of triplet excitons in OLEDs. Scientific Reports. 2015;**5**:8429. DOI: 10.1038/srep08429

[57] Ahn DH, Jeong JH, Song J, Lee JY, Kwon JH. Highly efficient deep blue fluorescent organic light-emitting diodes boosted by thermally activated delayed fluorescence sensitization. ACS Applied Materials & Interfaces. 2018;**10**(12):10246-10253. DOI: 10.1021/acsami.7b19030. https://pubs.acs.org/doi/10.1021/acsami.7b19030

[58] Byeon SY, Lee DR, Yook KS, Lee JY. Recent progress of singlet-exciton-harvesting fluorescent organic light-emitting diodes by energy transfer processes. Advanced Materials. 2019:1803714. DOI: 10.1002/adma.201803714

[59] Jang JS, Han SH, Choi HW, Yook KS, Lee JY. Molecular design of sensitizer to suppress efficiency loss mechanism in hyper-fluorescent organic light-emitting diodes. Organic Electronics. 2018;**59**:236-242. DOI: 10.1016/j.orgel.2018.05.015. https://www.sciencedirect.com/science/article/pii/S1566119918302374?via%3Dihub

[60] Kim HS, Park SR, Suh MC. Concentration quenching behavior of thermally activated delayed fluorescence in a solid film. Journal of Physical Chemistry C. 2017;**121**:13986-13997. DOI: 10.1021/acs.jpcc.7b02369

[61] Hatakeyama T, Shiren K, Nakajima K, Nomura S, Nakatsuka S, Kinoshita K, et al. Ultrapure blue thermally activated delayed fluorescence molecules: Efficient HOMO-LUMO separation by the multiple resonance effect. Advanced Materials. 2016;**28**:2777-2781. DOI: 10.1002/adma.201505491

[62] Kyulux Dramatically Improves the Performance of Its Blue Hyperfluorescence OLED Emitters. 2018. Available from: https://www.kyulux.com/kyulux/kyulux-dramatically-improves-the-performance-of-its-blue-hyperfluorescence-oled-emitters